Contractor's Survival Manual

by
William D. Mitchell

Craftsman Book Company
6058 Corte del Cedro P.O. Box 6500 Carlsbad, CA 92008

To Dianne, who said, you're going to write a what? But typed the manuscript anyway.
Thanks 123

Library of Congress Cataloging-in-Publication Data

Mitchell, William D.
Contractor's survival manual.

Includes index.
1. Contractors' operations--Handbooks, manuals, etc.
2. Building--Contracts and specifications--Handbooks, manuals, etc. I. Title.
TA210.M58 1986 624'.068 86-16620
ISBN 0-910460-42-6

Third printing 1989

Part One: Surviving

Chapter 1

Chapter 2

Chapter 3

Chapter 4

Chapter 5

Part One:

SURVIVING

- **Basics of Construction Contracting**
- **Which Way is Up?**
- **Finding Money and Finding Time**
- **Got That Sinking Feeling?**
- **Who's Minding the Store?**
- **Who, Me Work?**
- **The Equipment Payment's Past Due**
- **One Problem at a Time**

1 The Basics of Construction Contracting

This book is written for general contractors, builders and subcontractors, the self-employed entrepreneurs who handle most of the construction work in this country. Whether you're just getting started in construction or have been bidding jobs and meeting a payroll for years, you should find plenty of useful information between the covers of this book. Whether construction is your full-time occupation or your "other job" while you draw a salary on someone else's payroll, this book should help you make a better living in your chosen profession.

Construction contracting may be the quickest legal way I know to make money. Many contractors have doubled and redoubled their assets in a short period of time. It takes skill, luck, hard work and long hours, but the rewards are consistent with the risk and effort. Where else can you start out with a few tools, a truck and no special skills and build a multi-million dollar business in three or four years? It's been done many times in construction.

Even if you don't make a mint, construction contracting is satisfying work. You work outside, have only yourself for a boss, and can take pride in providing durable and attractive shelter, one of the most basic human needs.

But construction contracting is also complex and demanding work. Even a simple project requires coordination of many tradesmen and many different types of materials. Running that project (or a construction company) is like driving a team of spirited horses that wants to go off in all directions at once. Anyone successful at construction contracting is likely to be a jack-of-all-trades and the master of most, sort of a man for all seasons. A builder has to be a salesman, accountant, collection agent, labor negotiator, planner, plumber, laborer, estimator, marriage counselor and carpenter all rolled into one. If you come home tired at night, it's no wonder. Just one or two of these jobs would be enough for most people.

This Book Can Help

Because you're taking the trouble to read these pages, it's safe to assume that you have the problems that plague many contractors: You're not making enough money as a builder and you're frequently knee-deep in unpaid bills. Taking the time to read this book shows that you're one step ahead of the competition. You recognize the problem and are looking for a solution. That's an important step in the right direction.

This book is intended to help you sort out the complexity of running a contracting company. It's a guide, a road map to operating a successful construction business. It will suggest ways to get out of trouble, if that's where you are now, and explain how to build the more profitable construction company that you would like to have.

The first part of this book explains how to hang in there with what you've got. Survival comes first. There's no point discussing profits if your carpenters didn't get paid last week.

If your business is already doing reasonably well but isn't making enough money, concentrate on the second part of this book — thriving in the construction industry.

But in either case, I recommend that you read this book from cover to cover. It should help you maintain composure while making plans for the business you want to have.

Every bit of information in this book comes from my experience and personal observation. I know first-hand that the methods here work. But only you can judge if they will work for you.

The Author's Perspective

Before going any further, I'll describe the way I do business so you can understand my viewpoint. I'm a self-employed architect-construction manager by profession. In other words, I make my living designing and constructing buildings, just like you do. I've found a niche that's both profitable and comfortable for me.

I should point out, however, that finding that niche wasn't an easy or cheap discovery. As a builder, I've had several people working for me at some times, and at other times it's been just yours truly. Sometimes I made money, but many times I didn't. It seemed that every time I got rolling along, a recession came and took me with it. So I recession-proofed myself. I took a part-time teaching position. The money's not great, but it's steady and I make a lot of contacts. Next, I got rid of my staff, and did everything myself. Then I quit bidding, because there's too big a crowd at the bid openings. Finally I began to build one of two ways: I build what I design on contract for the owner with no competition, or I build projects that are so exotic there's no competition — everyone else is afraid of losing his shirt. I do these jobs also on a contract basis.

I never bid anything or compete for a job. I always know what I'm going to make going into a deal. All my help, except my attorney and accountant, are subcontractors. I don't carry anybody, ever. I always make money—more of it than I ever did as the typical, bid-on-everything contractor.

I'm not suggesting that my way of making a living is the only way. It's fine for me, but others may do as well or better taking the jobs I reject. The point I want to make is this: There are more ways to make money in the construction business than you can possibly imagine. And there are probably just as many ways to get wiped out. Competitive bidding (against cut-throat competition) isn't the only way to get work. It probably isn't even the best way.

Many construction contractors and subcontractors have found a profitable niche like I have — and do very nicely with it year after year. Maybe you can too. That you're reading this book tells me that you want to try.

Should You be a Construction Contractor?

Before going any further, ask yourself one critical question, "Should I be in the contracting business?" Does that sound like a dumb question? Maybe it is. But let's get the answer out in the open. How did you get into this business? Did you drift into it because nothing else interested you? Did you back into it because your father or uncle was a builder and you spent summers working for him? Or did you become a contractor after some thought and planning?

If you didn't give it any thought before, maybe now's a good time. Construction contracting isn't for everyone. The best framer or finish carpenter in the world may make a very poor construction contractor. There's more to construction contracting than just construction skill. If you can't estimate costs, sell the job, get a loan, maintain the books, collect bills, and keep your clients, employees and building inspector happy, your carpentry skill won't make you a carpentry contractor. Maybe you would be better off working full time as a carpenter and leaving the office work to others.

Ask yourself some more questions. Do you like accounting? Can you at least tolerate it? Everyone who runs a business, creates and uses financial records. If you hate paperwork, you'll never like contracting.

Can you work harmoniously with clients, associates, subcontractors and employees? Construction is a "people-oriented" business. If dealing with others isn't your strong point, maybe contracting isn't your best opportunity. You'll probably be happier doing something else.

Contractors also need a high tolerance to stress and a certain air of confidence and authority. I think of my job as being like the lion tamer at the circus. I work with some pretty independent and aggressive characters in situations where there's a substantial risk of loss. A mistake in a careless moment can lead to disaster. But if I've rehearsed my act carefully, there won't be too many surprises. And I've noticed that giving an impression of confidence and authority promotes cooperation and compliance in those I work with. If you don't thrive on stress and don't convey an appearance of confidence and authority, maybe contracting isn't for you.

As a contractor you can't take the demands of others too seriously. There are few real life-and-death situations in a construction business. No one is going to shoot you or lock you up for being in debt or losing money. In fact, you can beat or delay any creditor except the tax collector. Losing your nerve will just encourage those who want to take advantage of your misfortune. There's no need to panic . . . ever.

Your Attitude

It may seem irrelevant to worry about attitude when you're fighting for your economic life or struggling for more profits. But I can tell you from my experience that unless you believe without question that you'll succeed no matter what, you'll never get very far in the building business.

Your attitude influences how your clients view and judge you. If you come across as a negative, overburdened, marginally successful builder, you'll get only the left-overs . . . the work other contractors have turned down.

Property owners who have the money and borrowing capacity to put up a building or develop a plot of ground usually have something in common. I've found that they're successful, productive, practical people with a positive outlook on life. They like to work with other successful, productive, practical people with a similar attitude. They want to deal with contractors they trust, understand and like. And they're reluctant to work with those who seem to have different values and standards.

But don't misunderstand what I'm saying. I didn't suggest that you have to be a millionaire or have the look of a millionaire to do construction work for successful people. Many characteristics can offset a lack of financial resources. And prime among these are two that cost nothing, a positive attitude and self-confidence.

Accept the fact that your attitude is an important part of your business. It's the first thing your client sees and the last thing he'll forget long after his project is finished and sold. Let a positive attitude be a major asset of your company, even if the financial assets are a little skimpy.

Communication

Good communication habits are essential for every contractor. Some contractors ignore their clients, creditors and problems whenever possible. That's very poor policy. Problems don't go away by themselves. They require attention, consideration, and resolution. Above all, they require communication.

If you ignore your clients' phone calls and correspondence, you leave them no alternative but to come find you. Ignoring a client's call creates a problem in itself, even if there wasn't really any problem there before. You may think you're buying time by not returning a call. But the right attitude and good communication can win months or even years of delay . . . especially from creditors. So don't ignore your clients. Talk to them. You may have many unpleasant discussions, and you'll need a thick skin, but the alternative may be even more unpleasant, time-consuming and costly.

A Word About Profits

Understand this: Volume alone doesn't produce profits. At least 10% of all construction work is a potential money loser. When you find that 10%, turn it down and walk away. Low-profit work wears you down without producing an acceptable return.

You always work three times for a profit: once to find the job, once to do the job, and once to collect what you've earned. Just because you've performed the first two parts, don't assume that the third follows automatically. It doesn't. You have to protect, nurture, and collect profits before they reach your checking account. Otherwise you'll never take them home.

Profit-making is like popping corn. For best results, don't take your eye off the pot. Keep it in motion from the time the fat hits the fire until the last kernel bursts. Keep your job moving the same way. Watch your profit constantly. Nurture and protect that profit until work is complete. Profit is the only reason you have for doing the job. Don't forget that.

FINANCIAL STATEMENT

NAME: JOHN Q. AVERAGE DATE: OCT. 31, 1985

NO.	ASSETS	VALUE	NO.	LIABILITIES	AMOUNT
1.	REAL ESTATE	$87,500	1.	VALLEY LUMBER	$27,880
2.	AUTO'S & EQUIP.	17,300	2.	ABCO PLUMBING	33,210
3.	INVENTORY	38,700	3.	EAST BAY SAVINGS	128,940
4.	INSURANCE	1,275	4.	SO. FEDERAL BANK	73,775
5.	RECEIVABLES	34,890	5.	I.R.S.	27,940
6.	CASH IN BANK	1,550	6.	JOHNSON ELECT.	33,110
7.	LOANS DUE	1,225	7.	PHILLIPS HEATING	44,520
8.	CONTRACTS	43,900	8.	CONTRACTS	23,500
9.	HOUSEHOLD (MISC.)	5,150	9.	CREDIT CARDS	5,345
10.	TOTAL ASSETS	$231,490	10.	TOTAL LIABILITIES	$398,220
11.	NET WORTH EQUALS $231,490 MINUS $398,220 = (-166,730)				

Financial statement
Figure 1-1

Don't take jobs for owners who want you to work for nothing. Tell your clients right up front what it costs to build and what your fees are. Be friendly, honest and firm on the issue of profit. Some will respect you for it. Others may be offended. You can survive and even thrive without them. Let them go.

Building a reputation as the lowest-priced contractor in town will only wear you out physically and emotionally. Pick and choose your projects. Do quality work for quality owners that want to rely on a reputable, competitive and competent builder.

Preparing a Financial Statement

Accounting and paperwork are as much a part of a contractor's job as blueprints and estimating. And the most basic accounting document is the financial statement. You need a new one every time you apply for a loan or get a performance bond.

A financial statement lists what's owned and what's owed. It shows at a glance the net worth of the subject business or individual. It's like a financial x-ray. In the hands of someone who can interpret that picture, it speaks volumes about the financial health of that person or business.

Most going businesses prepare a current financial statement every month. If that's too much trouble, every three months will do. But going without a financial statement for more than three months is like setting out on a cross-country trip without a map.

If you've never prepared a financial statement, stop right now and do it. Figure 1-1 shows a sample financial statement. There's no point in reading any further until you know your assets, liabilities and net worth.

Here's how to make up a financial statement.

Use a blank piece of lined 8½ by 11 paper. Write your name or your company name at the top center of the paper and put the date right below the name. On the top left side of the page, write the word *Assets*. Under this heading make a list, by type, of all the money and things of value you or your business owns. Start with current assets: cash in your checking or savings account, receivables, inventory, stock or bonds, the cash value of insurance, and any advance payments made before receiving goods or services. Opposite each category write the realistic present value of that current asset.

Below the current assets, list fixed assets by category and value. Land, equipment, furnishings, vehicles and buildings are fixed assets. At the bottom of this list write *Total Assets* and the total value of all assets listed.

Now on the top right side of the sheet, list your *liabilities*. Then list by category everything you or your business owes: mortgages, charge accounts, loan balances, anything received but not yet paid for, money owed to subcontractors, invoices still unpaid and the like. Below all the liabilities write *Total Liabilities* and total the figures in the liabilities section.

On the last line at the bottom of the page, write *Net Worth*. That's the total of all assets less all liabilities. It shows what you're worth. If the net worth is negative, you're what's known as "being in the hole." It's not a great position to be in. But knowing your net worth is an advantage, even if it's negative.

Coping with Recessions

Every construction contractor should understand that there's a cycle of construction activity. This cycle rewards those that can anticipate it and punishes those that can't. The construction cycle can make or break you. And for many it does both.

At the beginning of every upswing in construction activity a fresh new crop of eager young builders surge into the industry. They develop a house or two, sell them off at a nice profit, and then tackle larger projects, making more money and laying bigger plans. After three or four very profitable years, some of these builders are running big construction companies with millions of dollars in assets and several major projects under way. They probably attribute their success to hard work, skill and daring. They're right. But they were also in the right business at the right time. And good times don't last forever.

When recession comes, as surely it will, hard work, skill and daring count for little. The bank loans, heavy investment in materials, equipment, staff, overhead and projects that can't be sold become a crushing burden. Many builders fail and leave the business. Others can salvage enough to remain active, or at least stay open for business until the next upswing comes.

Economic recessions are here to stay. There's no reason to suspect that our economy will be better managed or that recessions will be less severe in the future than they have in the past. Accept the ups and downs in construction activity as an opportunity to improve your competitive position against other contractors. Plan to survive when others can't and thrive when others can only recover.

Exactly what is a recession? From a builder's standpoint, we're in a recession when construction activity is down. That's usually because owners can't borrow money or would have to pay interest rates that make borrowing unattractive. Nearly all construction work is done on borrowed money. When lenders stop lending, builders stop building. That's a recession.

To survive more than one cycle in construction, you have to anticipate the construction cycle. It isn't hard. Like the seasons, they occur at regular intervals. Anything that's predictable can be planned for. And planning is the only way to make your company recession-proof.

Later in this chapter I'll explain how planning can help you use the construction cycle to your advantage.

Plotting the Construction Cycle

You can't anticipate the construction cycle until you know how the cycle works. So let's look a little more carefully at the construction cycle (or business cycle as it's sometimes called). Figure 1-2 shows the normal business cycle of rising and falling business activity. We've smoothed out the cycle a little to help you recognize the various phases of the cycle.

There's a definite trend during each part of the cycle. And what's going to happen next is quite predictable. Timing is the only major unknown. In spite of the rough shape of the actual cycle, the overall shape of the curve is usually a smooth continuous slope from a peak of inflation to the bottom of the recession. Since the curve is smooth and predictable, you'll have little trouble planning major business decisions around it.

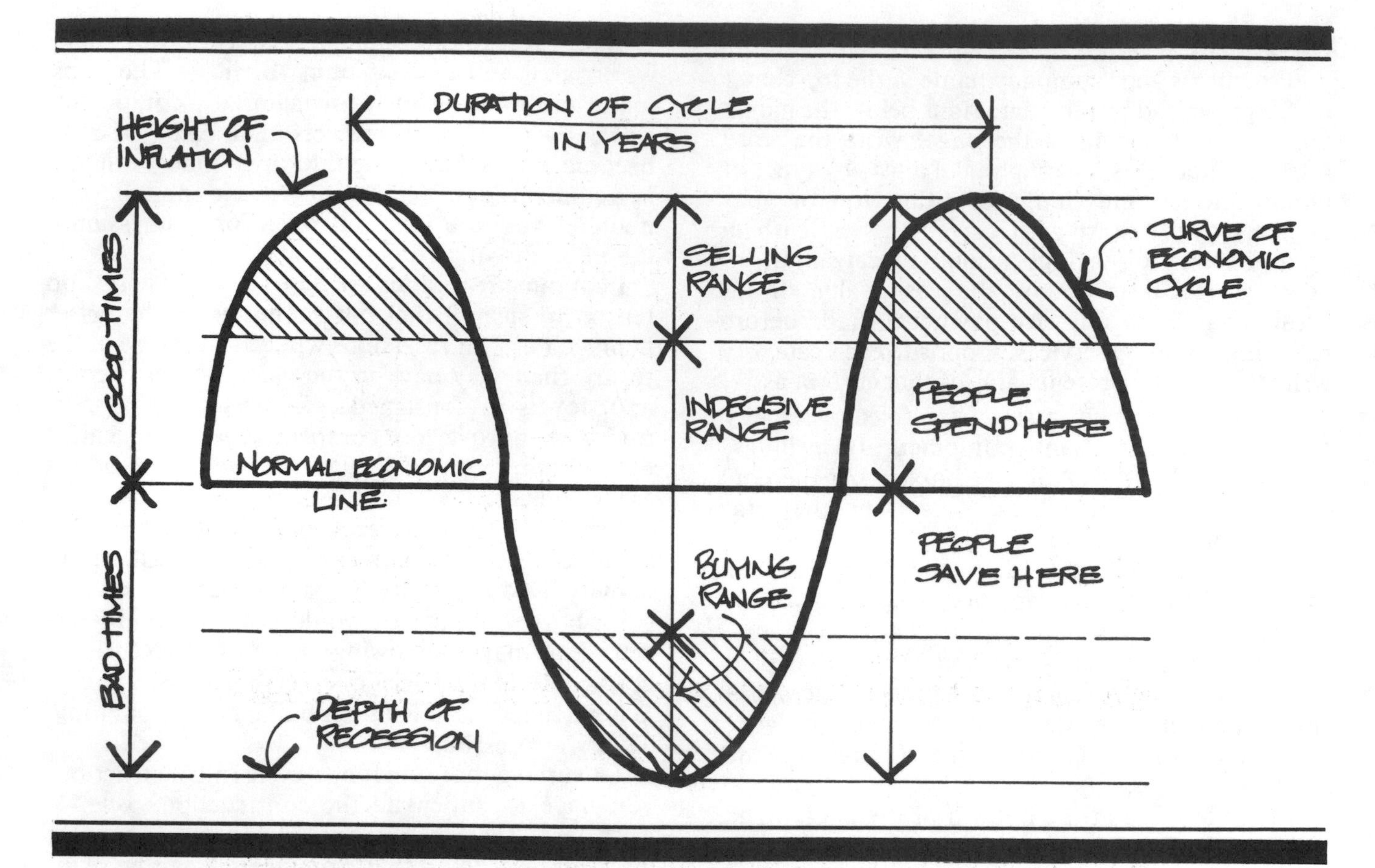

Normal business cycle
Figure 1-2

Don't worry about small fluctuations in the curve. It's only important that you identify the larger, slower, major changes in direction. See Figure 1-3.

Identify first whether we're in the recessionary or inflationary phase of the cycle. The small monthly fluctuations matter only to stock traders and commodity speculators. As a builder you're in a much longer-haul situation. You can't do much building in less than six months, so shorter-term fluctuations don't really affect you.

The overall trend in the economic cycle is generally upward or inflationary. For example, the lowest price of a home during the next recession will be higher than the lowest price of a similar home in the last recession. To get a better idea of what I mean, glance at Figure 1-4.

In Figure 1-4, you can see that the price of a home at the bottom of the recession in 1980 was $95,000. The price for a similar house at the bottom of the recession in 1974 was only $50,000. That's $45,000 less than 1980, representing a $45,000 inflation in values, even though it's a recession.

Notice that both the years 1974 and 1980 were deep recession years. Many land prices hit low points during those years. Yet we've noted a $45,000 increase in prices between those recessions.

The point of this explanation is to show that prices tend to increase even from recession to recession. Even if prices collapse completely in the next recession, count on them to rebound once more to new highs within a few years. That's just the nature of the business cycle. Knowing that should give you courage to hang on to assets when others are liquidating.

The real danger for builders and developers is that they'll be forced to liquidate at the bottom

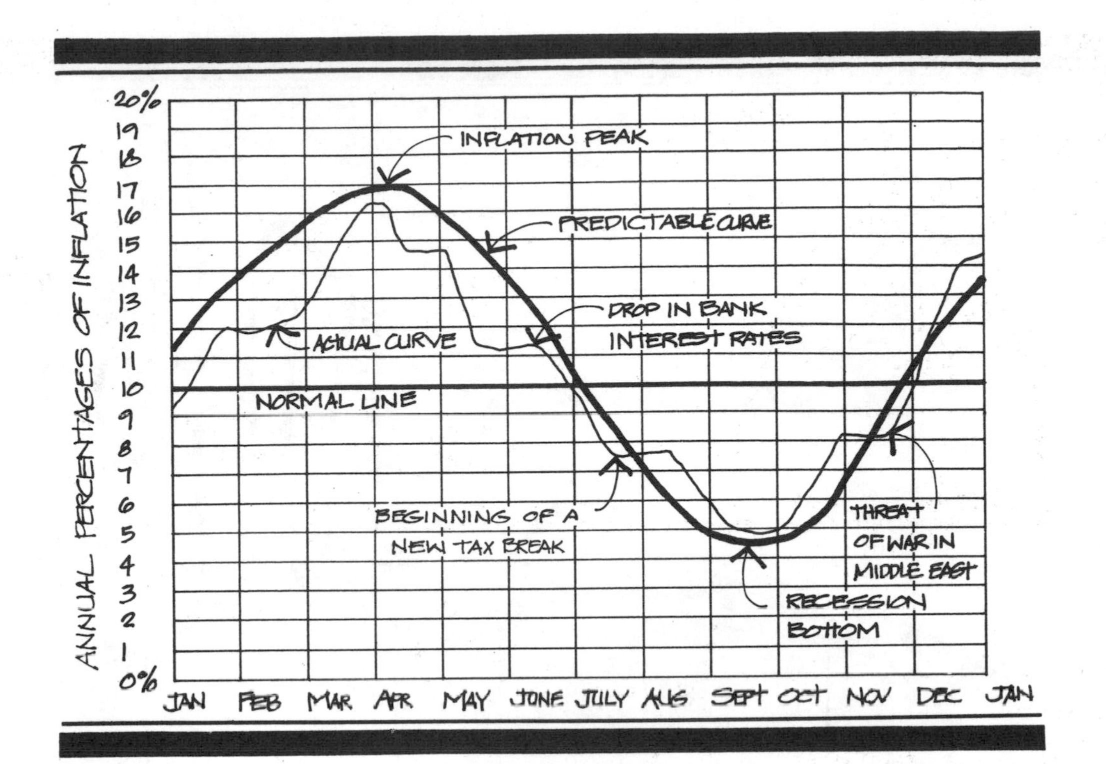

Actual inflation curve
Figure 1-3

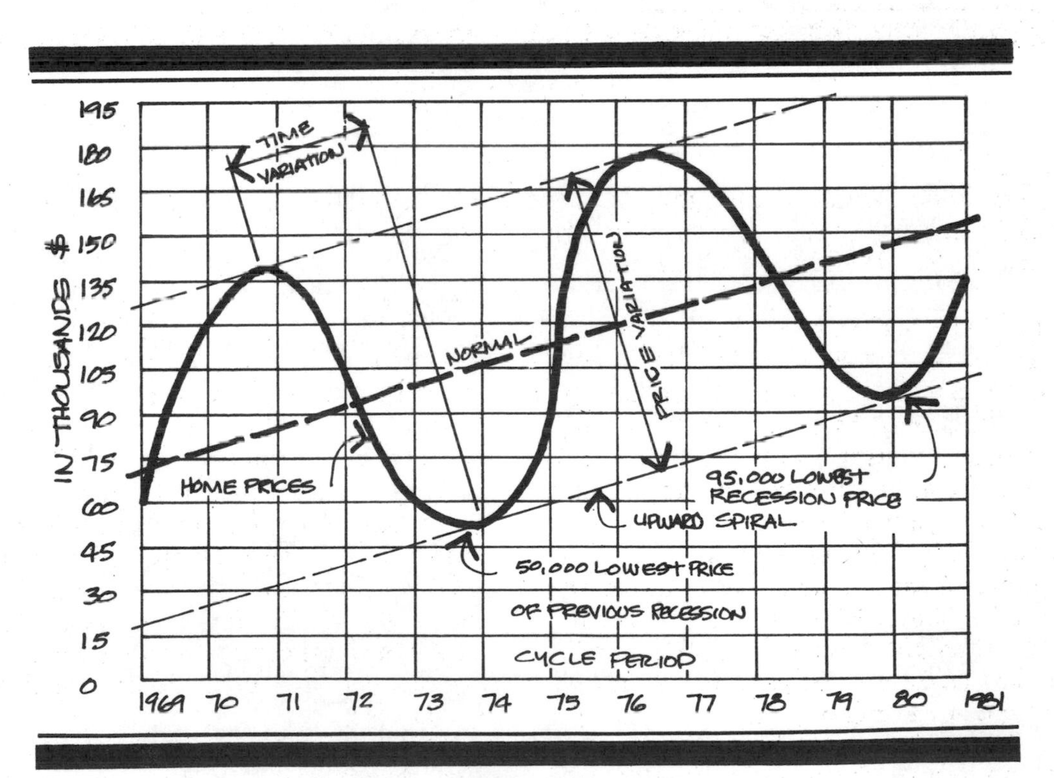

Upward inflationary spiral
Figure 1-4

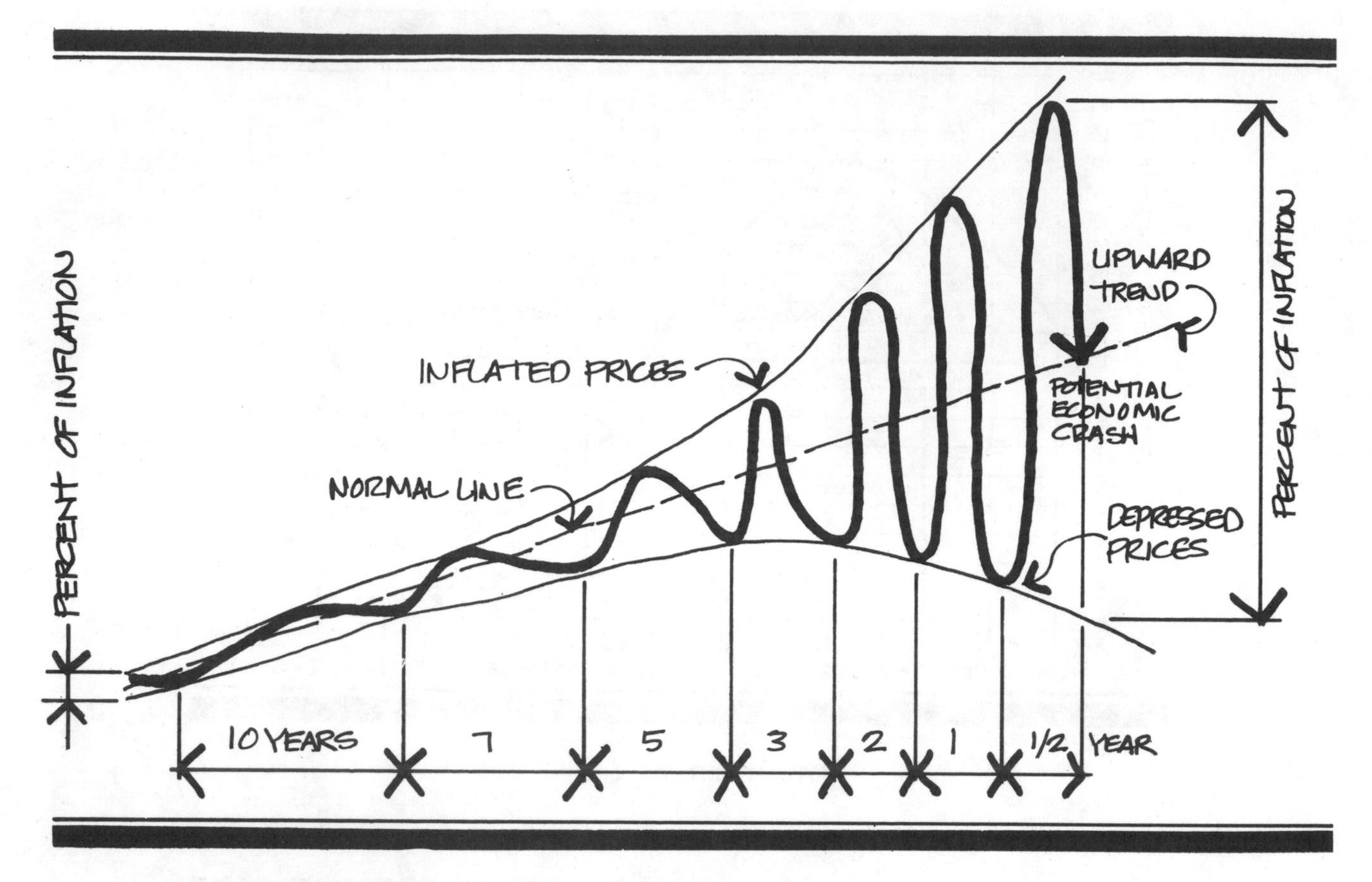

Spiral spacing
Figure 1-5

before the economy has recovered. Contractors that don't have the cash to make interest payments when due have to sacrifice assets to satisfy creditors or face foreclosure. A forced sale at the bottom of a recession is never good for the seller. And most important, it strips the contractor of the assets most likely to increase in value during the next upswing in the cycle. Have enough reserves to hold out during a recession and you'll never be forced to liquidate at fire sale prices.

Recessions — the Big Picture

Any discussion of recessions would be incomplete without a look at how often they happen. Let's look at recessions over a broader period of time to see if we can learn more than by examining a single cycle.

Look at Figure 1-5. Since World War II we've seen the inflation rate creep from an average of about 1 to 3% to a high of about 18%. Is inflation going to die any time soon? Not very likely. But don't let that bother you. From a historical standpoint, inflation is predictable. You can plan on it and use it to your advantage.

Notice that the inflation rate has tended to rise higher in each successive business cycle. That shouldn't be a surprise. What is surprising is the frequency and the steep slope during recent cycles. Figure 1-5 shows a distinct upward slope to the cycle over the years. Notice also that the peaks of the cycles are moving closer together. That's disturbing. Is it possible that the cycles could come so close together that you won't be able to tell the good times from the bad? As of this writing, it's too soon to tell. By the mid-1990's we should know for sure.

For now, just be aware that inflation drives prices upward in fits and starts, that the elapsed time between cycles seems to be growing shorter and that lately the cycles have become more exaggerated with each swing.

Planning for the Economic Cycle

So much for the way the cycle works. Now let's

take a look at the investment and business decision you should be making in each phase of the cycle.

How do you plan for this cycle? That's easy. Keep your thinking one-half step ahead of the construction cycle. Start thinking about the next recession when construction activity is intense. Then turn your attention to the next boom when recession is driving panic-stricken contractors to the wall.

When every carpenter who can drive a straight nail is working full time, begin thinking about what you'll do when the work in your shop is less than one-half present volume. How will you cut overhead by at least 50%? What projects will you close out as construction activity declines? Who are you going to lay off? What salary adjustment will you make? Can you shift emphasis to remodeling, additions or government jobs if more work is available there? Maybe you can get work on a "cost plus" basis at a slim but guaranteed profit while others battle it out for the lowest bid. Start building a financial cushion of spare cash . . . survival money to use when every dollar counts. For some builders it would be better to close up shop entirely for the duration of a recession. That's a real option. Leave it open. It may be better to close out your projects, pay your bills, furlough your employees and go back to teaching school or working for your Uncle Fred for a while.

Most of all, plan to reduce the risk of failure by reducing the money you owe. Only debtors end up in bankruptcy court. If you don't owe any money, you'll never go belly up. Debt-free builders don't have to take zero-profit work just to stay busy. They can put the business in mothballs or continue at very low levels of activity until better times return.

And better times *will* return. You know that, even when others have lost hope. Start planning your revival at the depths of the recession. When others are being forced out of the business, look for land or other opportunities that don't seem attractive at the time but have potential if buyers come back into the market. Start working with investors or lenders who can finance your growth during the boom. Decide what types of work you want to handle during the coming surge and prepare yourself and your organization for that day. Assemble your team and your resources for the next boom. Commit yourself and your finances as heavily as you dare to one key project or one opportunity that you feel will be most likely to succeed when construction revives. That's an excellent prescription for building a thriving business in the next revival.

Remember this throughout the construction cycle: Things are never quite as good and never nearly as bad as most people perceive them to be. Don't let the emotions of others keep you from using the construction cycle to good advantage.

For our purposes, we'll divide the cycle into three phases. See Figure 1-6. The top third we'll call the *inflationary period*. The bottom third is the *recessionary period*. The remaining or middle portion we'll call the *inconclusive period*.

There's a saying among developers that goes like this: "When people figure out what you're doing, it's time to switch to what they're doing." There's a simple truth here. Smart money moves in and out of the real estate and building markets as the cycle moves from inflation to recession and as the buying public reacts to inflation and recession.

The principle is so simple that most people miss it completely. If you plan to do next year what was generally accepted as good policy for last year, you're probably planning to do the wrong thing. There's a good time to invest and a good time to sell. If you buy when buying is considered wise by the general public, what you do is probably foolish.

And, strange as it may seem, there's a time to get out of the market altogether. It's not when everyone else is selling. When that happens, the really smart money has already sold out. The time to liquidate, or at least reduce your investment in real assets, is while there are still enthusiastic buyers left in the market.

When the recession is nearing bottom, when projects are being liquidated to satisfy creditors, smart money is picking up the choice assets that will be the first to recover when the economy revives, as it inevitably will.

The remaining part of the cycle, the middle third, is what we called the inconclusive period. Neither inflation nor recession is predominant. There is no clear trend or the trend may be in the process of reversing.

This inconclusive period is a good time to take stock of your position. It's a time to look seriously at what you've accumulated and weigh your options. It's a time of transition. It's time to restructure your thinking and reposition your assets from defensive to offensive, or vice versa. Few contractors recognize the need to make new business

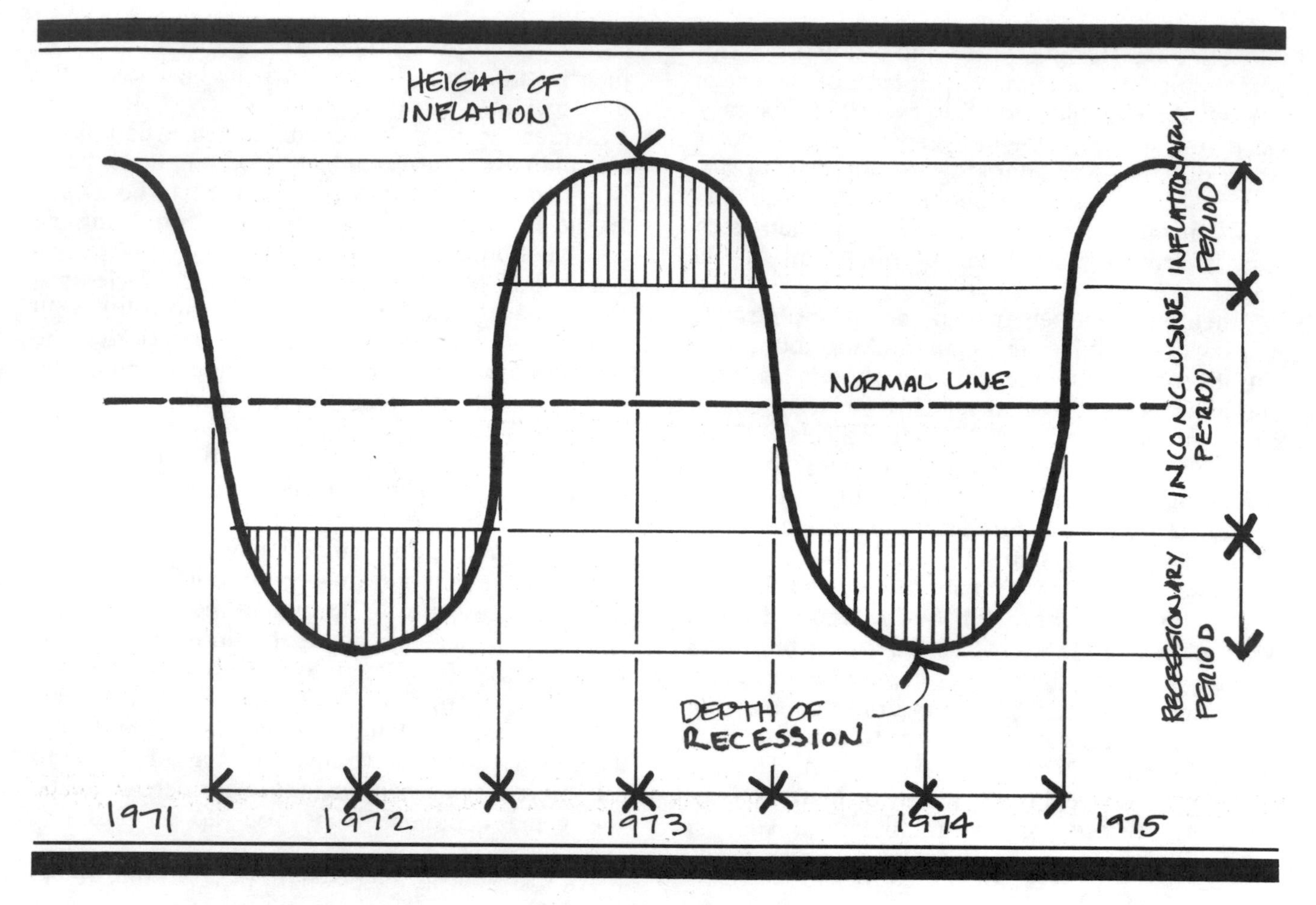

Inflation/recession periods
Figure 1-6

changes during the inconclusive period. Others see the need for change coming but don't make enough changes in time.

As I suggested earlier, the smart money moves in and out of the real estate and building markets. Think back to 1973. We were in a red-hot real estate market. Builders could do no wrong. What we in construction didn't realize was that we were cresting at the top of that particular inflationary cycle. By year end, Watergate had forced President Nixon to resign, credit disappeared, construction work stopped and the economy nosed over into a long recessionary slide that few were prepared for.

Look again at Figure 1-6. Notice that the economy had already begun to lose its momentum by the middle of 1973. This was a warning of what was to happen in the next few months. It was a sign to builders to conserve cash, postpone investment in additional equipment and real estate and to reduce staff until the next cycle began.

But few did. I watched builders, plumbers, suppliers and owners alike pushed to the brink of financial ruin. I laid off my staff of eight, could find very little work for months, and ran up $100,000 in debts. At the bottom, I had very little hope of working out from under these debts. No one knew how long the recession would last. Luckily, I was able to stick it out. I had to. There was no other way for me to pay off my debts.

Understand that timing is critical. No contractor can run his business as though the inflationary cycle will last forever. It never does. When the economy starts to crash, just get out of the way. You don't have to crash with it. Then lay your plans. Be ready to start new ventures when you're at the bottom of the recession, not the top. From there the prospects can only get better.

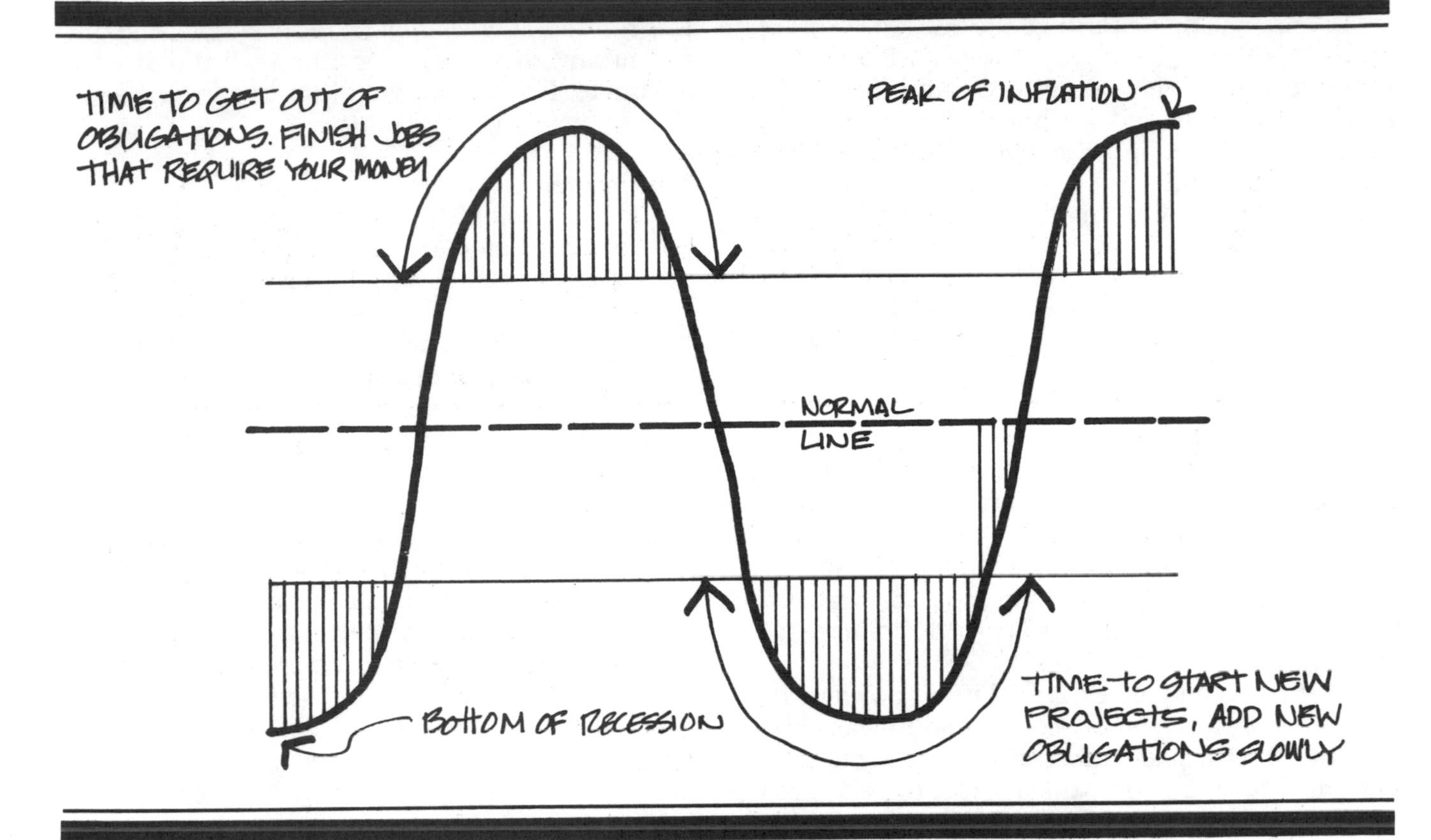

Obligation curve
Figure 1-7

But how do you know if the economy is at the top of a cycle or at the bottom? Unfortunately, it's not always easy. As I mentioned, the economic cycle is a rather rough curve, not a smooth sine wave as suggested in Figure 1-6. Even in retrospect, it may be hard to pick the exact bottom or exact top. Fortunately, that isn't necessary. It's only critical that you identify an inflationary or recessionary period when you're in it. Each period lasts at least 12 months. You have plenty of time to make up your mind.

In a typical four-year cycle, you have one year in an inflationary peak, one year of downward sliding through an inconclusive period, a one-year recessionary bottom, and finally a year of upward rising through an inconclusive period. The really critical periods are the top and bottom of the cycle. See Figure 1-7.

Let's say that you've been watching the economic trends. You feel pretty sure the cycle has peaked out and it's starting to head downhill into a recession. Now what? Well, it's time to wrap up any projects that are draining cash out of your pocket. Get out of debt now, before buyer enthusiasm is gone and bankers begin to get cautious.

Switching Horses

Switch horses before the inflationary bubble bursts. Stop building speculative houses. That market is about to collapse. Start bidding more work for other contractors. Go to work on someone else's money, not your own. If you wait until the recession's in full swing, the pickings will be slim. Too many contractors will be bidding for what little work there is available. Make the switch months ahead of when you actually need the work. Don't wait until six months after the need arrives. That's too late.

Suppose you're in the opposite situation. You feel the recession has just about bottomed out. Now what? It's time to begin those new projects you've been planning through the last 12 months of downsliding. Begin slowly though. There's no rush. Remember, it's better to get into the market three months late and get out three months early than to get trapped in a negative market with assets that can't be sold.

As the economy strengthens, put less emphasis on contract work with others. This gives you more time to devote to your own projects. But don't withdraw completely from the contract market. You'll need that work again in a few short years when the momentum of the current cycle is gone. Don't burn your bridges. You're going to need them again.

The kind of work to emphasize should be based on your evaluation of the economic conditions. Keep your workload in tune with the economic climate. If the economy is beginning to expand, gamble a little. Take short term risk. Speculate on a house or two, even a small tract or commercial building.

But if the economic cycle is shifting downward, it's time to switch horses. That doesn't mean closing down the business necessarily. Just avoid jobs that require a big investment. Go to work for others who are willing to risk their own cash, not yours. And do it before others recognize the trend in the economy.

To survive, strike a balance between speculative work and work you do for others. The quantity of each type should change as the economy changes. If you're doing 75% speculative work for yourself in inflationary times and 25% for others, reverse these percentages when the economy hits the skids. That maximizes your risk in good times and minimizes it in bad times.

This risk-taking isn't gambling. Gambling is tempting fate. It requires neither planning nor forethought. Risk-taking is a deliberately planned and carefully executed action. Take risks when the options have been fully explored, the possible losses evaluated, the time limits established and the money set aside. There's little, if anything, left to chance. Take risks when the odds are in your favor, like the dealer in blackjack. You may lose a hand or two, but over many hands, you're going to come out ahead.

By now you should see that trying to do business the same way in good times and bad can literally break you. Recognize economic cycles. Learn when to start your projects and when to close them out. You'll increase your profits and reduce the risk of loss substantially. Switch from speculative building to working on contract for others when that seems advisable. Believe me, you'll save yourself a lot of money, time and heartache.

Transferring Debt

Up to this point I've covered several important points that should be understood by every construction contractor. But I haven't done more than mention one of the key problems that most contractors have to live with: debt management. The rest of this chapter explains what to do when old bills can't be paid.

Let's say that we've hit the bottom of an economic cycle. In spite of all the planning you've done, losses on old jobs left piles of unpaid bills. There's money coming in on the current job. But it's not enough to clear up all debts on previous jobs. Here's your dilemma. Who gets paid? Suppliers and subs on your current job, creditors on prior jobs, the most insistent creditor, or a little here and there to keep everyone happy.

If you use receipts from current work to pay creditors on prior jobs, that's called transferring debt. There's one serious flaw in transferring debt. It doesn't work. You only succeed in creating a whole new group of unhappy creditors. Instead of having a single group of suppliers and subs you can't work with, you now have two groups. You create one more unhappy group of creditors each time you transfer the loss. Eventually word gets around that you sting everyone you work with. And no one wants to work with a slow pay contractor. So avoid transferring debt. It's notoriously unsuccessful and totally unrewarding.

There's only one kind of cash to transfer from one job to another — real profit left over after all bills have been paid. Otherwise, pay current debts first. You have to stay in business to pay off debts. Paying your current debts keeps you in business. If you go out of business, everybody loses — especially your oldest creditors.

You Can't Make it Up on Volume

Do you remember the story of two brothers who decided to go into the apple business? The first day in business they bought a truckload of apples at $10.00 a bushel and drove their truck to market. Unfortunately, apples were going for only $8.00 a bushel. But the brothers were in the apple business, so they sold off all their apples. Then they added up receipts for the day and discovered a problem.

The older of the two brothers sat down to think up a solution. After some time had passed, the older brother approached the younger one with his answer. "What we need," he said, "is a bigger truck."

The moral to the story is simple. If you're losing money, more volume will only lose it faster so you'll go broke sooner.

Don't fall into this trap. It's an illusion. If you're losing money, volume isn't the answer. It only compounds the problem. Sure, more volume may make more money. But *increased volume almost always requires accepting less profitable work.* That can be fatal for a contractor who's already in a profit squeeze.

Here's an example. Suppose you do a $100,000 job and make $10,000. You've made a 10% profit. Now let's say you do a $1,000,000 job and make $80,000. Sure, you've made more money, but the profit is down to 8%. While it usually takes more volume to make more money, every extra dollar in volume won't produce a proportionate increase in profit. And if you're not careful, eventually you'll catch one of those jobs that turns into a solid loser. Remember too that volume building requires more time and effort on your part and more staff, equipment and overhead.

Volume alone isn't important. Only profits are. If you're ever going to get out of debt, you need to make a profit on each and every job. The money allowed in your estimates for supervision will feed your family. The overhead allowance can keep your office running. But only your profits will pay off old bills. If you don't make profits, you won't get out of debt, no matter how many jobs you do. It's as simple as that.

Making money in the construction business is a lot like football. It's the short yardage plays that win games. It's consistency, day in and day out that wins, not the long shots.

Try to avoid the lure of the big job until you've got the resources: staff, equipment, cash and management. Grow into volume gradually. Don't create it instantly. In the construction business, ten years is no time at all. The big volume contractors are mostly third generation companies. How can you hope to compete with them? They've got a hundred-year head start on you.

Take my advice. Do what you do best and make a profit at it every time you do it. If you work for volume, eventually you'll get volume. But don't leap into a volume operation on the backs of unsuspecting suppliers and subcontractors.

Finding a Group and Regrouping

To dig out of debt, you must first regroup. By that I mean reorganize to cut expenses, even if it means going back to working with your hands for a while. If you're using the supervision money in your bid to pay a foreman, stop. His wage is money that could be used to feed your family. A foreman on the job is eating into your profits. You'll never pay off your debts that way. If you're paying a secretary and a bookkeeper, it's time to let them go. Have your wife answer the phone and do the books.

If you own expensive equipment, sell it and rent it back when you really need it. You're in no position to play super-contractor. High loan payments on idle equipment, unnecessary staff and extravagant overhead costs will bankrupt you. If you can't stand the thought of answering your own phone, driving an old truck and working on the job yourself, turn directly to Chapter 4, which deals with bankruptcy, because you're going to need it.

Regrouping is a slimming-down process. It's reorganizing to eliminate under-used labor and assets. It's also the establishing or re-establishing of a working group of professionals that can help salvage your business. This working group should include an attorney and a certified public accountant.

Here's my formula for regrouping a construction business on the brink of bankruptcy.

Start by evaluating the group of people that currently work with you. Review the performance of each. Look at what each of them costs you annually and what you're getting for your money.

If your attorney charges you for every phone call he places and each letter he writes, you have the wrong legal counsel for your kind of business. He may be O.K. for General Motors, but not for you. You'll go broke trying to pay him. Your attorney shouldn't be a family member or relative, either. Relatives are usually too close to the problem to be impartial, frank and realistic. They'll become "yes-men" if you let them. What you need is an attorney who has nothing to lose by telling you just how things really are.

It's not a bad idea to look for an attorney who may need your services as well. An attorney who's going to build a new home, an addition, or do some remodeling, may be willing to trade services. That helps you hold the line on legal expenses. But select a lawyer with the background and experience you need. My advice is to find a lawyer who specializes in real estate and construction work.

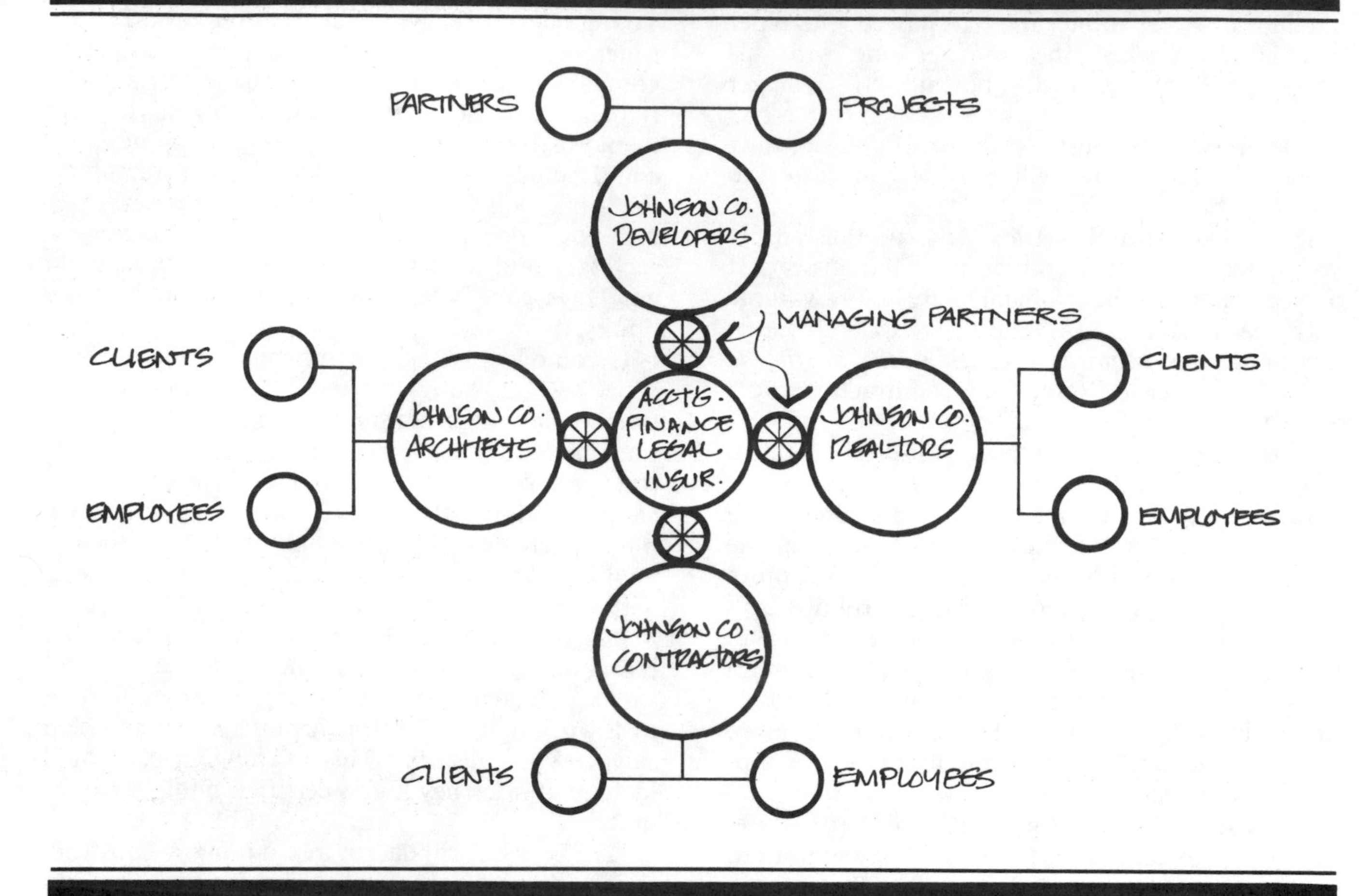

General organization
Figure 1-8

Follow the same rule when selecting an accountant. Certified public accountants are licensed by the state and are responsible for their errors and omissions. A bookkeeping service or unlicensed accountant usually costs less but may not be as reliable. Also, CPA's usually know more about tax laws, partnerships and corporate procedures.

Find professionals who are mature and stable. This may take time. But getting good legal and financial advice at a reasonable cost is worth the time and trouble.

When you regroup, make your organization as streamlined and frill-free as possible. Avoid computers, fancy phone answering equipment and the like. Don't hire help if you or a family member can do it and keep the money at home. That way you can always borrow it back if needed. If you're running several different companies, it's time to consolidate them into a single company. If you're operating as a corporation, consider the advantages of sole proprietorship. A corporation requires additional reporting and at least a couple of hundred dollars a year in corporate taxes. That money could be better used to pay off overdue bills. Figure 1-8 shows how the many functions of many businesses are organized around a single core of common services.

How Long In, How Long Out

How long will it take to get out of debt once the company has been streamlined and regrouped? Of course, it depends on how deep the debt is. There's

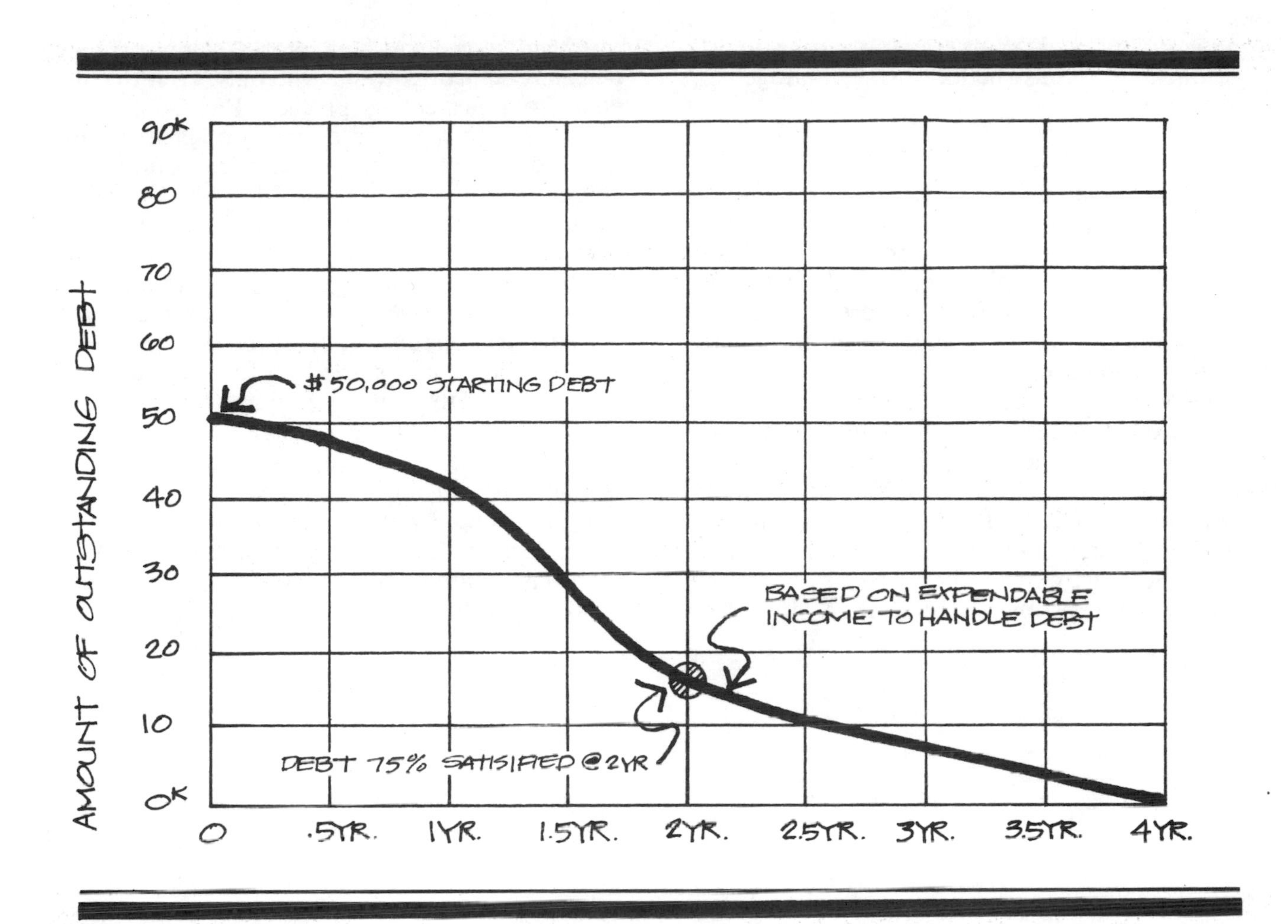

Debt reduction curve
Figure 1-9

no single answer. Some companies will never recover. But here's a rule of thumb that seems to work in many cases: It will take about twice as long to get out of debt as it took to get into debt. If two years of mistakes caused the problem, allow four years to recover. About 75% of your debts should be paid off in the first two years. The remaining 25% will require another two years. See Figure 1-9.

Paying off debts takes longer than creating them because a company deeply in debt has to reduce volume drastically to survive. Volume may be one-half or less the volume when times were good. During the inflationary phase of the construction cycle, it's easy to make money. But builders don't go broke during good times. They buy assets that they expect to sell or use. Then they go broke when work stops. That's when it's hard to make money and harder still to get free of debt.

If it took a year to get into debt, you'll spend the next year stalling, buying time, reorganizing and looking for profits. During that first year you'll pay off only 15 or 20% of your debt. In the second year you'll develop a steady cash flow that can begin to pay bigger chunks of your obligations, maybe 50 or 60% of the remaining debt. The last few creditors may have to wait for your third or fourth year of recovery.

Some creditors won't wait that long. But if they're going to sue, it will be during the first year of your recovery. I'll show you ways of handling lawsuits in Chapter 4. Remember also that 15 to 25% of all your debts will go away by themselves

— some of your creditors will simply write you off, or get out of the construction business altogether, or go broke themselves.

Summary

Contracting is demanding, complex work. There's so much that can go wrong. You have to wear a lot of hats and maintain a positive attitude. Remember that clients prefer "up" people and that communication is important. Don't ignore your clients. Talk to them. Tell your clients right up front what the costs will be and what your fees are. They'll respect you for it.

If you expect to survive in the building business, learn to adjust to the economic cycle. *You can't keep doing the same old thing in the same old way and expect to survive.* Be prepared for change. Study what's happening to the economy. Adapt your business activity to the times. Live on the money in your estimates for supervision. Use overhead money on your work to pay office expenses. Use profits to pay the bills. Don't transfer debt from project to project. Pay your current obligations first — and only out of profits. Don't fall into the volume trap. If you're losing money, volume won't cure it. Establish a working group. Include an attorney and a CPA. Don't play supercontractor. Trim your overhead: avoid the frills.

Once you've figured out the economic cycle, eliminated excessive overhead, reassembled your working group, finished the jobs that were losing money, and finally started to turn a profit, take some advice: Don't get carried away with yourself. Some contractors try to grow out of what they do best. Specialize in what makes money for you. Experiment if you must. But keep the experiments small and under control. Abandon what doesn't work. Get profitable and stay profitable. You'll soon have your money, self-respect and peace of mind well in hand.

Which Way Is Up?

"I'm going to sue you!"

If you owe money and can't pay it, you're going hear those words more than once. Of course, not everyone who threatens suit will actually sue. But probably one-third of your major creditors will sue within the first year after they discover you can't pay.

However, if you don't have any assets to lose, there's little reason to worry. Nobody can take what you don't have. The courts can't. Neither can they take away the way you make a living. If you're technically broke, relax. You're in the driver's seat. You own your creditors. All they can do is get a judgement that's impossible to satisfy. There's nothing they can levy against.

Your secured creditors will take back their security, of course. The bank will take the tractor they hold the paper on. Mortgage holders on land will get the land back. But that's all they'll get. As long as it isn't encumbered or in your name, it's hard for them to get at it. Generally they can't get more if the security is worth less than what you owe. And getting the security back may take a year or more.

Think of lawsuits as your creditors' last desperate gasp. They're trying to collect, and maybe punish you. Don't worry. Getting sued isn't so bad. Just keep your eye on the goal. What you want is to maintain as much control of events as possible. It's the loss of control that's scary, not the suit itself.

Here's an example of what I mean. You're sued by an irate creditor. Because you've got assets that you want to protect, you file an answer to the suit. That costs only about $200 to $300, and any attorney can do it for you. Even if you have no defense to the suit at all, you're at least a year away from trial. All during that time you remain in control. Your creditor can schedule depositions or make motions in court. But he probably won't because that's good money thrown after bad from his standpoint. And his attorney doesn't want to waste more time than necessary on this matter. Collection attorneys get paid a set percentage of what they collect, not by the hour. They play the odds. If you look like you're in such bad financial shape that the odds are heavily against their collecting, they won't waste their time on you.

There's going to be lots of delay in any suit. That leaves you free to go merrily about the business of surviving and laying the foundation for your recovery. You can sell assets, finish your jobs and start new projects. Meanwhile, your creditor is tied to the cumbersome legal system he decided to use. He can only wait for a year or two for the suit to come to trial. Meanwhile, you have use of his money. And that'll irritate him to no end.

INCOME INVENTORY

NAME: JOHN Q. AVERAGE DATE: OCT. 31, 1985

NO.	ITEM	SOURCE	AMOUNT
1.	CAL. STATE UNIVERSITY	BUSINESS CONTRACT	$12,100.00
2.	JONES RESIDENCE	BUSINESS CONTRACT	1,100.00
3.	ALAMEDA LOTS	SELL	18,000.00
4.	SIERRA PARK DISTRICT	BUSINESS CONTRACT	2,380.00
5.	COMPANY AUTO	SELL	800.00
6.	COMPANY TRUCK	SELL	6,500.00
7.	HOUSE	2ND MORTGAGE TRUST DEED	5,000.00
8.	MISC. BUSINESS EQUIP.	SELL	3,500.00
9.	JEWELRY	SELL	2,500.00
10.	PERSONAL AUTO	REFINANCE	1,750.00
11.	RENTAL HOUSE	SELL	5,000.00
12.	TOTAL INCOME INVENTORY		$58,630.00

Income inventory
Figure 2-1

Understand that I'm not advocating this. It's certainly not fair to the creditor. But it is, nevertheless, the way our legal system works. Try to remember this: bureaucracies move slowly. When a creditor elects to use the legal bureaucracy to solve his problems, he puts time on your side.

Debt Organization and Priorities

With your creditors stalled at least temporarily, let's turn our attention to organizing your debt load. Here's how we're going to do it.

Income inventory list— Start by listing your sources of income and assets. On a blank sheet of paper, list every source of income you can think of. Include in this list every dollar of profits, overhead, supervision and subcontract work you can pull out of every job you have under contract. Then add any stocks, bonds or notes owed to you. Add any equity in equipment, inventory and real estate you may own. Include everything of value. Refer to the financial statement that you prepared in Chapter 1 to make sure you've included everything. Figure 2-1 shows what this list might look like.

This income inventory list identifies your monthly income and the value of your assets. We're not going to use all these assets to satisfy creditors. But some could be sacrificed to keep you going. When

INCOME INVENTORY (ADJUSTED)

NAME: JOHN Q. AVERAGE DATE: OCT. 31, 1985

NO.	ITEM	SOURCE	AMOUNT
1.	CAL. STATE UNIVERSITY	BUSINESS CONTRACT	$12,100.00
2.	JONES RESIDENCE	BUSINESS CONTRACT	1,100.00
3.	SIERRA PARK DISTRICT	BUSINESS CONTRACT	2,380.00
4.	COMPANY AUTO	SELL	800.00
5.	HOUSE	2ND MORTGAGE-TRUST DEED	5,000.00
6.	MISC. BUSINESS EQUIP.	SELL	3,500.00
7.	JEWELRY	SELL	2,500.00
8.	PERSONAL AUTO	REFINANCE	1,750.00
9.	TOTAL INCOME INVENTORY (ADJUSTED)		$29,130.00

Income Inventory (adjusted)
Figure 2-2

you've made that decision, remove the items selected from your income inventory list, apply them to your debts, and retotal the column. What remains is what you have to work with. See Figure 2-2.

Indebtedness priorities list — Divide a second sheet of blank paper into three vertical columns. Column one is for essentials only. They'll be paid 100% each month. Column two creditors will get small monthly payments. Column three is your wish list . . . no payments at all for the present, just a big wish. Tell them their name's in the hat. If they object, you can always offer to take their name *out* of the hat.

In the first column, list your *minimum* monthly living expenses by type: food, shelter, clothing, utilities, health, transportation, education and incidentals. In the second column, list all of your small debts. This is anything under, say $1500, plus anything owed to suppliers necessary for daily business operations. In the third column, list all remaining debts: anything over $1500 that is owed to anyone that isn't essential to your recovery plan. These are debts that can't be paid from present income. See Figure 2-3.

Include in column three creditors that have already sued you or seem bent on suing. Candidates for this column include your more hostile and stubborn creditors. If you have large corporate creditors that are just following procedures in the company directive "How to Handle Past Due Accounts," put them in column three. Smaller creditors with a belligerent attitude also go in column three.

If any creditor in column two sues you, move him immediately to column three and stop paying. Replace him with someone more reasonable from column three, and begin monthly payments to *that* creditor.

INDEBTEDNESS PRIORITIES

NAME: JOHN Q. AVERAGE DATE: OCT. 31, 1985

NO.	LIVING EXPENSES		PAYABLE EXPENSES		UNPAYABLE EXPENSES	
1.	MORTGAGE	$547.00	MID BAY SUPPLY	$47.00	SIERRA SUPPLY	$5470.00
2.	FOOD	285.00	JOHNSON LUMBER	250.00	HAL'S PLUMBING	2780.00
3.	UTILITIES	175.00	ACME ELECTRIC	175.00	L&D CONCRETE	3990.00
4.	PHONE	70.00	J&T HARDWARE	68.00	JOHN HANSON	2150.00
5.	INSURANCE	110.00	BILL'S ROOFING	85.00	BUD'S REPAIR	1790.00
6.	MEDICAL	50.00	TAXES	75.00	JIM'S RENTALS	650.00
7.	CLOTHING	100.00	MISC.	35.00	T&M GUNITE	8760.00
8.	ENTERTAINMENT	50.00			VALLEY LUMBER	6180.00
9.	GAS/AUTO	265.00			BOB MURPHY	845.00
10.	MISC.	75.00			ELLEN'S SECY. SERV.	435.00
		$1747.00		$735.00		$33,050.00

Indebtedness priorities list
Figure 2-3

The overriding idea in debt organization is to eliminate as many small debts as possible as quickly as possible. That's column two. Column three will have to sue or wait until they're paid voluntarily. Ultimately this is the end of your relationship with them.

Be honest with your creditors. Tell them that you'll pay in full, each and every time they call. Be friendly, forthright and optimistic. Don't sound depressed. Shouting and threats won't help. This is strictly business. No moral issues are involved. But don't guarantee any specific time for payment. Just let them know that you have a list of creditors, that they're on that list, and that they will be paid as soon as money will stretch that far. But you have to feed your family first!

You'll repeat this same conversation over and over until your creditors become impatient and give up on arm twisting. Then they'll sue or assign your debt to a collection agency. This is exactly what you expected when you made up your indebtedness priorities list. You've been as honest as possible with each creditor. You intend to pay every creditor in column three and you told each one of them so. There's not much more you can do beyond that.

What you need is time, lots of time. Organizing your debts and priorities will buy that time. You may not like what you're going through, but a builder in deep financial trouble has no other choice. Remember, the whole point of the first half of this book is *survival.* If you don't survive, nobody gets paid. *Everybody* loses.

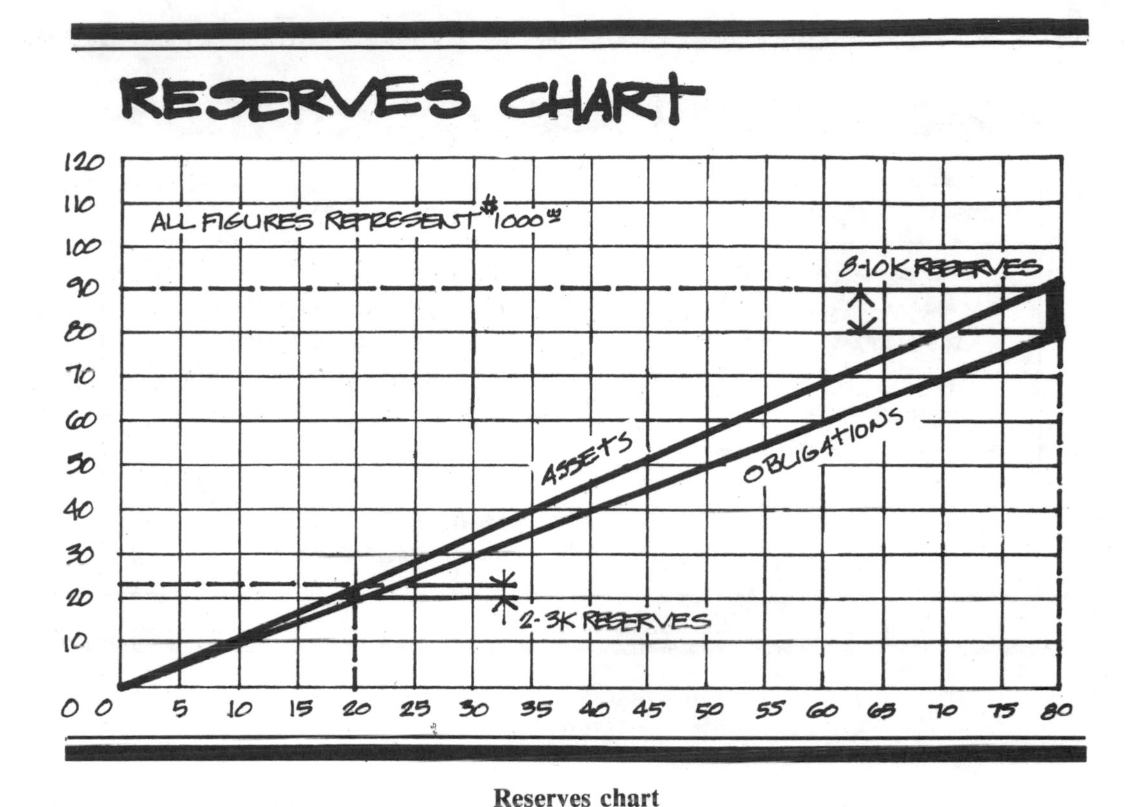

Reserves chart
Figure 2-4

Too Little Debt
Owing a creditor too little money can be a problem. Most states have a small claims court that lets creditors sue without an attorney and get a judgement in about a month. That can be dangerous for you. The sheriff could be camping at your door or drawing money out of your bank account about 60 days after you're sued. That makes small claims court a potent threat.

Fortunately, there's an upper limit on small claims court actions. It's about $1,000 in many states. That's why I recommend paying off the small debts and ignoring the larger ones for the present. An option a creditor does have if you owe him more than the limit for small claims actions, is to give up the portion over the limit, and sue for only the maximum amount. But if he doesn't want to give up the excess, he'll need an attorney and many months to collect. That's to your advantage. How much should you owe a creditor to force him out of small claims court? A good rule of thumb is 50 to 100% more than the maximum amount. Few people hesitate to lop 10% off a delinquent bill. But most won't give up 40 or 50%. After all, they worked hard for that money. They won't give it up without a fight.

Here's something to keep in mind, however. If you can stall a creditor by owing him more than the small claims limit, your debtors can do the same to you. No one has a monopoly on this information. It's available to anyone. *Consider setting up agreements so that final payments you receive are no larger than the small claims limit.* If a client stalls you on a final payment, get him into small claims court immediately and save the one-year wait.

Remember, most creditors are just like you — in debt. They're just in debt to different people. To work successfully with debt, you must keep it in perspective. Everybody owes money to someone. It's nothing to be ashamed of. But a heavy debt burden in relation to income is something to be genuinely concerned about.

Once you're back on your feet, take a solemn oath to maintain a balance between debts and assets. Debt shouldn't be more than 75 or 80% of assets. A bank would use about the same yardstick when making a loan on your house. Cash reserves

COLLECTION AGENCY ACTIVITIES

LEGAL ACTIVITIES	ILLEGAL ACTIVITIES
1. CAN CALL YOUR HOME. 2. CAN VISIT YOUR HOME. 3. CAN SEND YOU NOTICES. OR LETTERS AT HOME OR AT YOUR OFFICE. 4. CAN TALK WITH WIFE OR HUSBAND.	1. CAN'T HARASS (VERY VAGUE). 2. CAN'T CALL REPEATEDLY. 3. CAN'T CALL AFTER NORMAL BUSINESS HOURS 8AM-8PM. 4. CAN'T CALL WORK WITHOUT YOUR PERMISSION. 5. CAN'T VISIT OR TALK TO YOUR PLACE OF BUSINESS. 6. CAN'T CONTACT FRIENDS OR NEIGHBORS.

Collection agency activities
Figure 2-5

(including anything that can be turned into cash quickly) should be at least 10% of total assets. See Figure 2-4.

While battling to hold off creditors, remember that they are in the same predicament as you. They fight receivables every day to stay current on payables. They have to collect from you and others like you to survive. You also have to collect to survive. Collections are just part of a contractor's job. Get comfortable with collections. It comes with the territory.

Collection Agencies and Your Credit Rating

Here's how to handle collection agencies: ignore them. Until their attorney sues you, don't even give them a thought. If they sue, tie them up in court for a year or two. A collection agency has to follow the same procedures as any other litigant. They have no advantages and have one big disadvantage. They don't know the facts about your debt.

But you know one important thing when a collection agency comes calling. That means that the creditor has given up on the debt. He's assigned it for collection. The agency gets about a third to one-half of everything they collect. Who wins on that assignment? No one but the agency. They have to work only once for their money.

When you are sued and lose, an abstract of that judgement can be recorded at the county recorder's office. This isn't cleared until the debt is paid off. The abstract attaches as a lien to any real property you own in *that* county, but not in any other counties. Collection agencies and credit associations run a similar record-keeping system on debtors. A collection agency puts your name on file with local and national credit associations for seven years.

If a collection agency is operating within its legal rights, there's not much you can do but tolerate them. But the law limits contact with you to once a day. If a bill collector becomes belligerent or contacts you late at night, feel free to hang up or slam the door.

Ignore telegrams and "24-hour" written notices. They're effective only against unseasoned debtors who take them seriously. The collection agency's greatest weapon is harassment. It's the squeaky wheel that gets the grease. That's their philosophy in a nutshell. Don't let yourself be intimidated. Be courteous and candid, as long as they treat you with respect.

The law regulates what collection agencies can and can't do. Most of what they *can* do is listed in Figure 2-5.

CREDIT RATINGS

CREDIT RATINGS	DESCRIPTION	PREFIX DEFINED
R0	TOO NEW TO RATE	R= REVOLV'G. ACCT.
R1	AS AGREED	I= INSTALLMENT ACCT.
R2	30 DAYS PAST DUE	O= OPEN, 30, 60, 90
R3	60 DAYS PAST DUE	DAY ACCT.
R4	90 DAYS PAST DUE	NOTE: THE LETTERS
R5	CHAPTER 13 BANKRUPTCY	R, I & O CAN BE
R6	REPOSSESSION, VOL. OR UNVOL.	INTERCHANGED TO
R7	COLLECTION, CHARGE OFF, SKIPPED	SHOW TYPE OF ACCT.

Credit ratings
Figure 2-6

For a modest fee, you can get a copy of all the information in your credit file. This information is not available to the general public. You have to give creditors permission to review this data. That's something you do routinely nearly every time you fill out a credit application.

Credit ratings affect your life in several ways. Let's take a look at the common rating systems. Usually the letters RO, R1, R2, etc. are used to indicate a degree of creditworthiness. See Figure 2-6.

Your credit rating can vary from year to year as your ability to meet obligations changes. If you've developed a poor credit rating, consider placing an explanation in your credit file. That can be done at any time. But you should do it as soon as you've turned the tide of your financial difficulties. Just the dry facts about your credit problems can look pretty bleak if there's no explanation or suggestion that you're on the road to recovery.

But there's hope, even if your credit report looks very grim. A funny thing happens when you have money in a bank. Maintain a respectable balance, stay current on your obligations to that bank, and your credit rating doesn't mean a thing when it comes time to get a loan. As everyone knows, money talks. If you have cash in the bank, suddenly your credit history with that bank is as clean as new-fallen snow.

Trading Services and Working Off Debts

Trading services is nothing more than the old-fashioned barter system. But it has one outstanding advantage: The Internal Revenue Service has a hard time taxing income and profits on services received through barter.

Barter isn't as convenient as using cash. But barter works when people lose confidence in cash or when cash is scarce. The trouble with barter is that it's hard to find someone who both needs what you've got and can offer what you need. But if you're buried in past due accounts, you have a ready-made list of barter prospects, your creditors. These people have needs. What can you offer them?

Take your dentist, for example. You owe him for some work done last year. Could he be looking for a new microwave oven? If so, you buy it at wholesale and resell it to him at the discounted cost. The discount probably covers your bill.

The dentist may take one additional step. He may deduct your dental work from his taxable income by taking your bill as a bad debt. That's not

VALUE DETERMINATION

NO.	COMPANY	EST. VALUE	LOANS LIENS	EQUITY
1.	EASTSIDE REALTY	$54,500	$37,500	$17,000
2.	JONES, SMITH AND ANDERSON	61,000	"	23,500
3.	NATIONAL CITY REALTY	47,500	"	10,000
4.	WHITE, MARTIN AND ASSOC.	39,500	"	2,000
5.	BOWIE RIVER REALTY	65,500	"	28,000
6.	ALTON STREET INVESTMENTS	52,000	"	14,500
7.	AVG. VALUE IS $320,000 ÷ 6 = $53,333	$320,000	$37,500	$95,000 ÷ 6 = 15,833 EQ.

Value determination
Figure 2-7

legal, of course. But no doubt it happens. The IRS would consider your trade a taxable event — both you and the dentist received something of value. But the dentist's records show no payment, and the debt isn't added to your credit rating.

You may be able to liquidate or reduce debts by remodeling, adding on, or by building new facilities for a major creditor. Do you have a boat, a mountain cabin or tickets to a ballgame? Negotiate debt reductions with your creditors. Do some estimating, buy an item for a friend of a creditor or provide some service for a creditor of your own creditor.

The possibilities are limitless. It doesn't always require cash. All it takes is a little imagination, and the guts to ask, "What do you need besides money?" Trading services for debt really works. And it's certainly preferable to a law suit.

Liquidating Assets, Quitclaims and Trust Deeds

Let's say you've organized your debts, battled the collection agencies into submission, and set up trades with several creditors. Now it's time to start liquidating the assets you identified as expendable earlier in this chapter.

There are several ways to sell assets. If you have real property assets that require monthly mortgage payments, look at them first. They may actually be a liability more than an asset. From a cash-flow point of view, it may be in your best interest to liquidate them as quickly as possible. But before you unload this or any type of asset, determine how much equity you have in the property.

Refer to Figure 2-7. A quick way to appraise real estate is to ask half a dozen local realtors to give you an opinion of the value. They'll usually do it free if there's a chance that they'll get the listing if it's put up for sale. The average of these six opinions is probably close to the fair market value. Now deduct all loans and any liens on the property. The remainder is your equity. If there's little or no equity left in your asset, unload it.

If the equity is nearly zero, consider using a quitclaim deed to dispose of the property. A quitclaim deed is a valid deed. But it makes no promise or warranty about who actually owned the property.

Quitclaim Deed

This Indenture *made the ...., ... day of*

.. one thousand nine hundred and ....

Between

the part...... of the first part,

and

the part...... of the second part,

Witnesseth: *That the said part...... of the first part, in consideration of the sum of*

.. dollars,

lawful money of the United States of America, to .... in hand paid by the part...... of the second part, the receipt whereof is hereby acknowledged, do .... hereby release and forever QUITCLAIM unto the part...... of the second part, and to .... heirs and assigns, all th...... certain lot....., piece......, or parcel...... of land situate in the ....

.. County of ....

State of .... , and bounded and described as follows, to-wit:

Together *with the tenements, hereditaments, and appurtenances thereunto belonging or appertaining, and the reversion and reversions, remainder and remainders, rents, issues, and profits thereof.*

To Have and to Hold *the said premises, together with the appurtenances, unto the part..... of the second part, and to .... heirs and assigns forever.*

In Witness Whereof *the part...... of the first part ha...... executed this conveyance the day and year first above written.*

Signed and Delivered in the Presence of

.... } ....

.... } ....

This document is only a general form which may be proper for use in simple transactions and in no way acts, or is intended to act, as a substitute for the advice of an attorney. The publisher does not make any warranty, either express or implied, as to the legal validity of any provision or the suitability of these forms in any specific transaction.

Quitclaim deed
Figure 2-8

FULL EQUITY LOANS

1. ORIGINAL APPRAISED VALUE $100,000.00
2. FIRST TRUST DEED LOAN 80% OF VALUE 80,000.00
3. HARD MONEY EQUITY 20,000.00
4. NEW APPRAISED VALUE 120,000.00
5. SECOND TRUST DEED LOAN 80% OF VALUE 96,000.00 – 80,000.00 EQUALS .... 16,000.00 (80,000.00 IS AMOUNT OF 1st TRUST DEED LOAN)
6. PRIVATE LENDER THIRD TRUST DEED LOAN OF 4000.00 EQUAL TO BALANCE OF EQUITY, BASED ON INFLATED APPRAISAL. 4,000.00
7. TOTAL OF 2ND & 3RD TRUST DEED LOANS EQUAL 20,000.00 WHICH IS THE AMOUNT OF REMAINING EQUITY IN THE PROPERTY. CONSEQUENTLY IF YOU HAVE NO EQUITY YOU HAVE NO OUT-OF-POCKET RISK.

Full equity loans
Figure 2-9

It only transfers whatever rights the seller had. If he had none, nothing is transferred. The deed is usually filed by the seller at the county recorder's office.

Generally, filing a quitclaim deed relieves you, the seller, of any financial obligation connected with the property. If you choose to use a quitclaim deed to eliminate an obligation, use a document like Figure 2-8.

If you have a substantial equity in your property, you'll need a willing buyer who can pay a good price. But before selling, consider giving yourself a "buy-back" option in the sale. This gives you cash now, when you need it most, but preserves your right to buy the property back on a certain date at a set price plus interest. Most buyers will balk at an option like this. But it's going to work very well with friends and relatives who would like to help you out but want the security of an asset to protect their investment.

If the equity is large enough, consider taking another loan on the property. It's commonly used by homeowners. A home is a very negotiable asset. Banks and S&L's make good money lending on single-family, owner-occupied homes.

Although the distinction is breaking down, banks tend to make different types of loans than savings and loan associations. Banks generally deal in short-term loans, such as automobile and home appliances loans. Savings and loan associations make property loans, but fewer short-term installment loans.

When a financial institution makes a loan, it takes back a security interest in some property that has more value than the amount borrowed. In some states this is called a mortgage. Others refer to the instrument as a deed of trust. The effect is the same. The lender gets the legal right to take the property if the borrower defaults.

The mortgage or deed of trust is entered at the county recorder's office and becomes a public record. The first instrument recorded is the most senior and has rights against all instruments recorded later. Later lenders on the same property can take title to the property only by paying off lenders who recorded their deeds of trust or mortgages

100% FINANCING

1.	ACTUAL TRUE VALUE OF YOUR PROPERTY.	$100,000.00
2.	125% APPRAISAL OF YOUR PROPERTY.	125,000.00
3.	NEW 80% 1st or 2nd T.D. OF $125,000.00 APPRAISAL. . . .	100,000.00
4.	MINUS ORIGINAL 1st T.D. FROM FIGURE 2-9.	80,000.00
5.	HARD MONEY TO YOU FROM FIGURE 2-9.	20,000.00
6.	NEW EQUITY REMAINING IN YOUR PROPERTY. . .	-0-

100% financing
Figure 2-10

earlier. That makes the first lender more secure than later lenders.

If you're financially solvent, it's usually easy to get a first or second mortgage up to 80% of the value of the property. But you'll fight an uphill battle to get a third mortgage from banks and S&L's.

The holder of a third mortgage is in a precarious position. If the borrower defaults, first and second mortgage holders will get paid from the proceeds of sale. But if the property value has dropped 10 or 20%, due to a sacrifice sale, the third mortgage holder is wiped out. He gets little or nothing.

To the contractor who's on the ropes, a third trust deed may be as good as a sale. The full 100% of value has been taken out of the property. See Figure 2-9.

Remember that lenders make their loans on opinions of value. That's usually the opinion of a certified appraiser. It's extremely difficult to get a bank or savings and loan association to lend beyond a set percentage of appraised value. 75 or 80% is the usual figure for first mortgages. They won't adjust their appraised value. But they may increase the percentage to 85 or 90% if you can make a good case that the loan will be trouble-free and profitable for them.

Your creditworthiness makes a difference when taking out any loan. If you foresee a loan in the future, get it before your financial problems become acute. Do it before news of your credit problems reach the local credit agencies.

Remember, when you borrow money to pay your bills, you're not reducing debt. You're just swapping creditors. Every swap costs you money for an appraisal, title search, loan fees, and document recording, not to mention the interest. Don't add to your debt burden unless absolutely necessary. And when you borrow, have some predictable income source that will meet the monthly loan payments. Otherwise, all you're doing is trading creditors. There's no advantage to that. Worse, you run the risk of adding one more black mark to your credit file.

Occasionally you'll get an appraisal that is either way too low or way too high. Neither is a problem. If it's low, just go to another lender and get another appraisal. Appraisals at 125% of the actual value do occasionally happen. If it happens to you, you'll be able to borrow to about 100% of the market value either by refinancing the first mortgage or by adding a new second mortgage. See Figure 2-10.

Hard Money Loans and Factoring

When you've borrowed as much as banks and S&L's will lend, it's time to turn to private lenders. The rates tend to get higher, but the requirements are more flexible. Some private lenders will lend close to 100% of the assumed value. Some will lend

3RD T.D. 100% FINANCING

1.	ACTUAL TRUE VALUE & APPRAISED VALUE OF YOUR PROPERTY	$ 100,000.00
2.	NEW COMBINED 1ST & OR 2ND T.D. LOAN FROM BANK @ 80% OF APPRAISED VALUE	80,000.00
3.	ORIGINAL 1ST T.D. LOAN FROM FIGURE 2-9 TO BE PAID OFF	80,000.00
4.	ORIGINAL EQUITY (HARD MONEY) UNRECOVERED	20,000.00
5.	NEW 3RD TRUST DEED LOAN FROM PRIVATE LENDER	20,000.00
6.	NEW EQUITY REMAINING IN YOUR PROPERTY	-0-

3rd T.D. 100% financing
Figure 2-11

over 100%, either through misjudgement in evaluating the property or as a gamble that the value will inflate quickly. They make the loan knowing full-well that you may ultimately default on the loan. Then they can relieve you of the property. I call this the *vulture syndrome*. See Figure 2-11.

Quick loans from private lenders on property you already own are called *hard money* loans. That distinguishes them from *soft money* loans, which are notes sellers take back when selling property. In a hard money loan, the lender has to actually put cash in the hands of the borrower. In a soft money loan, the transaction is normally just a paper one.

To find private lenders, ask realtors who are active in your community. The classified section of your local paper probably includes ads placed by investors who want to buy second mortgages and by property owners who need to raise cash. If those sources aren't fruitful, call on a bank loan officer, mortgage broker or finance company. They'll know the private lenders who are active in your community.

Private loans usually have a short fuse. Even though the loan may be amortized over a ten-year period, it's probably due and payable in three years. And the rate will usually be the highest allowed by law. You can see that these loans are only for the desperate and those who think they can make a killing with the proceeds. The payoff can be very difficult to handle, especially if your financial position doesn't improve quickly enough.

Most hard money loans are repaid by taking another loan from an institutional lender when the first loan matures. That refinancing almost always requires a loan against some other real property that has available equity.

Hard money loans can be helpful if you have absolutely no other alternative. The loan buys time but adds another set of entanglements. Remember, borrowing to pay bills isn't a way out of debt. It's just substituting one creditor for another.

Factoring, on the other hand, is different. A lender, or factor, buys your receivables for cash. Your creditors then make payments to the factor. You get cash up front as soon as the account is due.

Most major banks do some factoring of receivables. If you've spent a considerable amount of time with one bank or another, that's probably a good place to start looking for a willing factor.

Factoring can solve cash flow problems if you are growing quickly and have heavy receivables for a relatively short period. The cost of factoring is heavy. But it may be the only alternative if there are no other assets available. But don't use it to make up for losses or to carry excessive overhead.

Factoring is no picnic. I've never seen a builder use it successfully for any length of time. Most contractors don't have heavy receivables from solvent debtors. But if you do, here's how the factoring system works. You assign receivables that are less than 30 days old to your factor. The factor deducts his fee from the invoice and sends you a check for the balance. The fee is usually 10 to 50% of the amount due. The older the receivables, the smaller the amount, the less likely the account is to pay, the greater will be the factor's fee.

Summary

The law is a funny thing. It lets any creditor chase you relentlessly but makes the course long and treacherous. And if necessary, the law lets you beat creditors out of their money through bankruptcy. You can't change the legal system. But you can use defects in the system to your advantage if you need to. Take the time to make up an income inventory list and an indebtedness priorities list. These valuable tools clear your thinking on your debts and income. Remember, the objective in debt organization is to eliminate as many small debts as possible, as quickly as possible. Don't be intimidated by collection agencies. They have to use courts like everyone else. And don't overlook the opportunity to trade services for debt. Liquidating your assets, borrowing against real estate, and factoring, are other options available to you. As a last resort, go to a hard money lender.

I've emphasized several times that creditors are not your adversaries. Think of them as disappointed friends. They were there to help when you needed their labor and materials. You're the problem. You couldn't pay the bill when their work was done. It's not their fault that you underestimated the job they worked on, or overspent your share of the job's income. Naturally your creditors are upset. They're not mad at you personally. They're just disappointed with your business performance. You've created the image of a poor business manager, and perhaps you deserve it.

But images change. Deal honestly with your creditors. You're going to pay them off. When that happens, they can say you were slow to pay, but they can't say you were dishonest.

Why bother to deal fairly with creditors? Two reasons. First, you're laying the groundwork for financial recovery. That's much easier if you preserve your reputation in the construction community. Everyone admires someone who manages to struggle back from insolvency if he eventually pays all his creditors. Abraham Lincoln did it. So did Harry Truman. And both were later elected President.

But there's a more important reason. You're going to come out of this with your self-respect and dignity intact. Recovering financially is going to make you a much wiser and more capable contractor. My recovery did that for me. It will for you too. It's inevitable. The most important benefit from working your way out of debt is what it does for you.

You went into debt of your own free will. Now, by standing your ground, you're going to get out of debt. Fight your way back to solvency. That's the first step. The second is building a profitable, financially-sound construction company. Think of the first step, recovery, as the training ground for the second. That's why you plan to pay every creditor and meet every obligation.

Finding Money and Finding Time

Borrowing doesn't pay bills. It only buys time — at a price. Interest is the price you pay for the time you need. Figure 3-1 shows that price. It's an interest rate chart. Let's say you wish to borrow $40,000 in order to stave off some creditors who've been badgering you. If you borrow at 10%, you'll increase your total debt by $4,000 annually, until you've paid off the loan.

You've paid your old creditors, but now you've got an upset banker on your hands because you can't make the loan payments on time. Also, you're $4,000 deeper in debt than you were before. And that's if you were able to borrow at a modest 10% interest rate. Look at the chart again. At 15%, you're going to be $6,000 deeper in debt, and at 20% you'll owe a whopping $8,000 more. This loan wasn't a good idea.

Figure 3-2 shows how debt compounds. If you have $200,000 in old debts and you borrow $100,000 at 20% to pay off half of them, you've just created $20,000 in new debt, shown shaded in Figure 3-2. Now let's take this newly created $20,000 debt and project it over the five-year loan life. You could end up with $100,000 more debt than you started with, if you can't keep current on your interest payments. Also not a good idea.

If you have to borrow, you should know how to do it right. That's the purpose of this chapter — borrowing from creditors, both willing and reluctant. Let's start with the obvious: interest rates.

Some lenders and sellers deliberately make it hard to compare interest rates. They use computation methods that make the rate they're quoting seem more competitive than it really is. Fortunately, there's a single, uniform standard that's prescribed by federal law. It's the annual percentage rate (A.P.R.). Don't put too much faith in interest rate quotes that omit those three key letters. Once you have the A.P.R., it's an easy matter to compare loan rates.

If you have to borrow to buy time, you'll need some documents: A current financial statement (profit and loss and balance sheet if you're self-employed) and your tax returns for the previous two years. The bank will supply its own loan application form.

It's against the law to submit phony documents when applying for a loan. Don't even think about doing it. But it's completely legitimate to supply documents that put you in the most favorable light without actually mis-stating a fact. Your accountant will be able to suggest some legitimate ways to dress up your assets and de-emphasize your liabilities. But don't carry this puffing too far. Lenders are quick to spot inflated assets and understated liabilities.

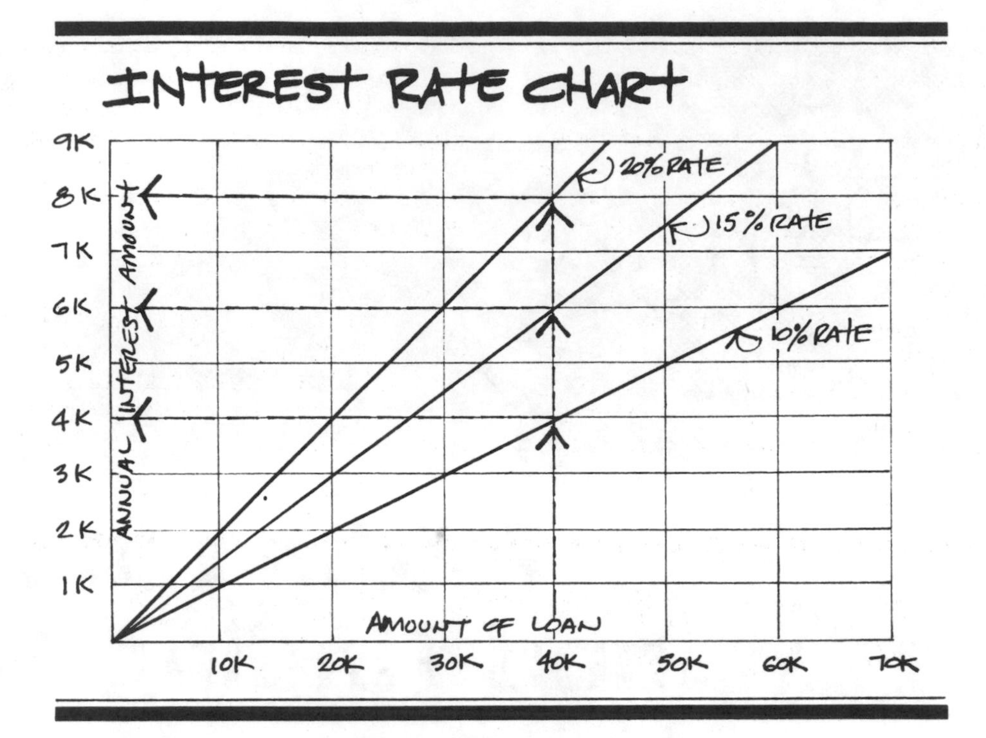

Interest rate chart
Figure 3-1

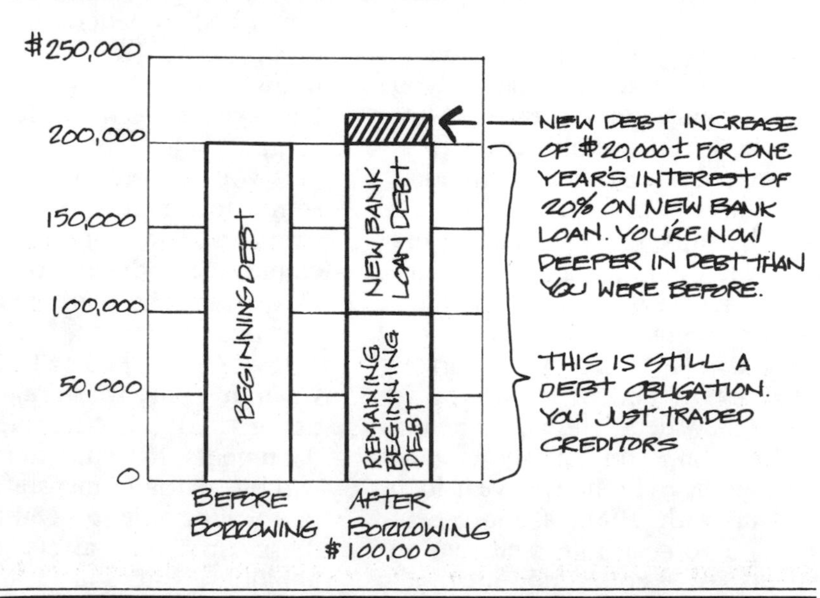

Interest and compounding debt
Figure 3-2

Review the adjusted financial statement in Figure 3-3. Compare it to the original financial statement, Figure 1-1 in Chapter 1. The point of adjusting the financial statement is to avoid showing debts that aren't really due yet, or debts that won't show up when researched. If possible, head off new billings from creditors until you've completed your new loan processing. Your creditors will probably be glad to cooperate, since your loan gives them some assurance they'll get paid.

Include in your assets all jewelry, stocks, bonds, business net worth, tools and equipment. Generally, you can exclude from your financial statement loans from relatives and friends that are moral obligations rather than legally enforceable debts. But be sure to include all secured debts. These are recorded with the state or county where you live and will show up in a credit check.

You'll notice that in adjusting the statement from Figure 1-1, I've dropped off some debts and added some assets to get the financial statement in Figure 3-3. I've increased my net worth as much as possible.

The tax returns you submit with your loan application are assumed to be the same returns you filed with the federal and state governments. But there's no way for the bank to check that. They have to take your word for it. I wouldn't advise it,

ADJUSTED FINANCIAL STATEMENT FOR LENDER

NAME: JOHN Q. AVERAGE — DATE: Oct. 31, 1985

	ASSETS	AMOUNT		LIABILITIES	AMOUNT
1.	REAL ESTATE	$187,500	1.	VALLEY LUMBER	$25,110
2.	AUTOS & EQUIP.	34,850	2.	ABCO PLUMBING	27,585
3.	INVENTORY	52,700	3.	EASTBAY SAVINGS	128,940
4.	INSURANCE	1,275	4.	SO. FED. BANK	73,775
5.	RECEIVABLES	54,795	5.	IRS	27,940
6.	CASH & LOANS	13,470	6.	JOHNSON ELECT.	19,870
7.	CONTRACTS	171,985	7.	PHILLIPS HEAT'G.	35,980
8.	OFFICE FURN.	16,335	8.	CONTRACTS	15,475
9.	HOUSEHOLD	18,770	9.	CREDIT CARDS	2,780
10.	TOTAL ASSETS	$551,680	10.	TOTAL LIABILITIES	$357,455

NET WORTH $194,225

Adjusted financial statement for lender
Figure 3-3

but some contractors have been known to prepare one return for filing and another for loan applications. The second return shows the highest possible income that can be justified. They increase their taxable income by minimizing deductions and including every penny of cash income from overtime or small jobs done for cash.

Who decides whether you get the money? As you might guess, it's a committee. All banks and S&L's have loan committees that decide who gets loans and who doesn't. The committee's role is to protect depositors and shareholders in the institution, and allocate money among creditworthy borrowers. But you'll never meet with the loan committee. At least, I never have. Instead, you'll work with a loan officer whose role is to prepare the paperwork and make a recommendation on the loan.

Some loan officers will suggest what the loan committee is looking for in financial statements and tax returns. They know what the committee will approve and what the committee will scrutinize closely. If your loan officer suggests ways to put your application in the best possible light, follow that advice. The loan officer doesn't want to waste time on applications that can't be approved any more than you do.

Have your accountant prepare the company balance sheet and profit and loss statement. That way they appear to come from a neutral party. But the statement doesn't have to include an auditor's opinion. That's a waste of time and money unless you're running a publicly-held company.

Be sure to give the lender the most current information possible. Data more than six months old is considered obsolete.

Figure 3-4 shows a sample profit and loss statement. When you're making up a profit and loss statement for a lender, carefully consider both income and expenses. Put your financial picture in its best possible light. As in your financial statement, you'll list only your trackable expenses while including all deals that produced income in the form of cash, checks or assets. You want to show a big income and a small outgo.

Buying Time, Stalling and Floating

Your recovery takes time. Getting enough time may require the use of every asset at your disposal. A major asset you can't afford to overlook is "plastic money," the credit cards in your pocket. Think of the credit cards as prearranged lines of credit you can draw on when needed.

You can make cash loans against most major credit cards. Use this money to pay past-due construction bills. The more credit cards you have, the more credit you can draw on. Stay within the credit limit on each card and make the minimum monthly payment on each card for as long as possible. But don't use this credit if money is available from other sources. The balance due on your credit card will carry an interest rate that's among the highest charged for all loans.

If you've reached the credit limit on each card and can't make the minimum monthly payment, it's time to begin stalling. That's a simple maneuver. You put off an obligation for an undetermined length of time with a promise. It involves no cash payment on your part.

Promise your creditors that your condition is only temporary. You assure them that it's the result of events beyond your control. Use every excuse that seems appropriate, but be as truthful as you can. After all, you want to come out of this with your dignity and self-respect intact.

If all else fails, tell the unvarnished truth: You're a poor business manager. Explain that you understand the creditor's problem and are sympathetic. You know he deserves to get paid, and you wish you had the money to pay him. But you don't. You will certainly understand if he decides to sue. But if he does, you'll transfer him to column three of your creditor list and stop all payments. Explain that you can stall him legally for years because you have no other options available. Explain that if enough creditors won't cooperate, you'll be forced into involuntary bankruptcy. Then everybody will lose.

Chapter Thirteen of the bankruptcy laws lets you reorganize your debts and put your creditors on ice for three years. If that doesn't work out, you can file for Chapter Eleven bankruptcy, which discharges nearly all unsecured debts completely. We'll talk more about bankruptcy in the next chapter of this book.

Straight talk will persuade many creditors. They'll take your advice and wait. Others will take a tougher course and sue. Either way, you get more time to recover.

Another option to consider is your *float*, checks you've written which haven't yet cleared your bank. Nearly all contractors use their float at one time or another. You write a check at the lumberyard, knowing full well that there isn't enough money in your account to cover the check.

ITEM NO.	DESCRIPTION			RUNNING BALANCE
		BEGINNING BALANCE		$15,743
1.	JAN. TO MAR.	INCOME	87,441	18,501
		EXPENSES	68,940	
2.	APR. TO JUNE	INCOME	103,912	16,419
		EXPENSES	87,493	
3.	JULY TO SEPT.	INCOME	127,775	17,923
		EXPENSES	109,852	
4.	OCT. TO DEC.	INCOME	73,556	6,073
		EXPENSES	67,483	
	TOTAL PROFIT OR LOSS (ANNUAL)			$74,659

ADJUSTED PROFIT AND LOSS STATEMENT FOR LENDER
COMPANY NAME: JOHN Q. AVERAGE CO. DATE: 1.1.85

Adjusted profit and loss statement for lender
Figure 3-4

But you also know that there *will be* enough tomorrow morning when you deposit the check from the Jones job.

Banks help you compound the float. They credit the payee's account when a check is deposited rather than when funds are actually withdrawn from the payor's account. Here's an example. A California contractor deposits a New York check in his California account. He gets credit immediately at his California bank. But the funds aren't withdrawn from the New York account until probably three days later. The check has to be flown back to New York before a debit is made to the New York account. Both the California contractor and his New York client have use of the same money for three or four days.

Banks are getting quicker at clearing checks. And some restrict use of deposits made with out-of-state checks. But until they reduce the float to zero, there's an opportunity to use money that doesn't really exist. That's an important opportunity when you're strapped for cash. Float will buy only a few days at best. But sometimes those can be very important days.

You probably see another possibility for creating float already. Suppose the transaction isn't between you and a client. Suppose your account is on both ends of the transaction. That's called check *kiting* and it's illegal. Here's an example of kiting stretched to its maximum:

Even though there's only $10 in your *personal* account in Bank 1, you write a check on that account for $5,000. The check is payable to your construction company. You deposit the check to your *company* account at Bank 2. Then you wait. You're going to let this check bounce before you make a deposit to Bank 1.

One note of caution here. Make sure these are different banks and not just different branches of the same bank. Even better, make sure the banks are separated geographically as far as possible. Checks clear too fast when the payee and the payor use the same bank or different banks in the same city. Deposit the check after three o'clock and postdate the check if possible. That adds more delay to processing the check.

It takes two or three working days for the inter-bank system to process your hot check. Let's say that your check was processed in three days and has just bounced. Now it's on the return trip through the bank processing circuit back to Bank 1. This time allow only two days in the system. Five days have gone by. For five days you had an extra $5,000 in your company checking account.

On the fifth day you write a check on your company account in Bank 2 for $5025 and deposit it to your personal account at Bank 1. The extra $25 is to cover the bounced check charge. When Bank 2 calls to report that the first check has bounced, you assure them that the check is now good and instruct them to clear the check again. Three days later the first check clears on the deposit of the second check. If there's no money in Bank 2, the second check is now in the inter-bank circuit and will bounce in about three days.

Let the check at Bank 2 bounce once before you do something about it. When it bounces, you have to either write a check on a third bank or come up with the cash to end the whole process.

This is no way to make a living. And it's illegal. If you never make a real deposit, you're going to end up explaining it all to the sheriff. In some states it's a felony if the check exceeds a certain amount.

Some kiting schemes go on for years and involve hundreds of thousands of dollars before they're discovered. But banks are getting better at detecting kiting and are less reluctant to testify against their customers that try it. My advice? Don't abuse the relationship you have with your bank. There's too much that they can do for you.

Dealing with Your Banker

There's no substitute for an understanding banker. A banker who's on your side is an invaluable asset in many ways. Banking is a game of trust. You plan for the bad times in the good times. You develop a friendship with your banker when everything's coming up roses. During the good times you borrow a little and pay it back. The amount is up to you. If security is required, use your house, some equipment or a vehicle. Borrow enough so that it involves your banker if you run into trouble. You want him to feel that he has a stake in your survival.

The first time you can't meet a payment, keep your banker fully informed. In Chapter 1, I explained why communication with your clients is very important. This goes double for your banker.

If you have an unsecured loan with a bank and you're behind on the payments, offer to convert the note to a secured loan with your car, home, receivables or an assignment of sale proceeds. This will extend the time for payment for months, or even years.

If you're behind in your mortgage payments and the bank is ready to foreclose, arrange for a quick sale to someone you know well who is in better financial condition than you are. This will stall the bank for the three or four-month escrow period. Assign to the bank the sale proceeds. Usually your buyer can qualify for the loan and take title with almost no money down.

If you have equipment payments or loans that are past due, you can usually "roll" notes over. By that I mean that you can pay only the interest on the loan and sign a new note with a new payment schedule and due date. Basically, you're starting over from scratch on that particular obligation. But the bank isn't obligated to do this. They do it as a favor to you, their good customer.

There's another privilege that many banks grant to their good customers. Banks need deposits to make loans. If they don't make loans, they can't survive. A customer who keeps high four-figure or low five-figure balances in a checking account is a good customer. No bank would close that customer's account just because there was an occasional overdraft. Neither are they going to bounce that customer's checks. Some contractors with heavy cash flow get away with murder with overdrafts on their checking accounts. No bank would boast about something like this, but it happens.

Cash Flow, Conserving Income and Avoiding Debt

Contractors live and die by their cash flow, the money coming in every day. Even if every penny that comes in has to be paid out again the same day, a construction company can survive for years in apparent prosperity. But that isn't unique. A family, a partnership, a corporation, a city, a state or even a nation can operate quite successfully even though fundamentally bankrupt — as long as the amounts of money coming in and going out are roughly equivalent. Under conditions like this, even minor fluctuations in cash flow may cause panic. But it may not cause the house of cards to collapse.

Yet builders with a cash flow of a million dollars a month have survived with liabilities equal to their assets. So long as each dollar of outgo is balanced by a dollar of income, they're doing fine. But if income is one dollar and outlay is one dollar and ten cents, there's a problem. That's negative cash flow.

Let's say that you have an income of $10,000 a month and expenses of $9,000 a month. Your profit before tax is $1,000. But your business needs a new rig that costs $50,000. Your plan is to pay for it out of profits for the next fifty months. That's four years and two months of going without a profit, assuming that the new rig doesn't improve your profitability.

You go ahead and make the purchase. A little while later business drops off to $7,500 per month and remains there for the next two years. That's the normal construction cycle, as explained in Chapter 1. Expenses remain at $9,000 a month and you still have to pay $1,000 a month for the new rig. At the end of the two years you're in the hole by an additional $60,000.

By some miracle, you're still in business when the construction cycle swings upward again. The new equipment is nearly one-half paid off. Income is back to $10,000 a month and expenses are still at $9,000 a month. But now you have $60,000 in additional debt. Interest on that debt is more that $500 a month, leaving only $500 a month available for payments on the rig. So you refinance the equipment to lower payments to $500 a month for the next 50 months.

Unfortunately, we must assume that prosperity won't last that long. It seldom does. Long before that rig is paid off, cash flow will drop off again and another loan will be needed. Even if the rig is eventually paid off, it will take you five years to dig out from under the $60,000 debt load.

Dependable cash flow is a big asset to any contractor. Unfortunately, very few contractors have it. Fluctuating cash flow can be disastrous in an undercapitalized business. The more limited your cash reserves, the more important it is to keep income and expense in proportion.

If your expenses are roughly the same every month, even a slight dip in income may be difficult to handle. If income drops for several months, you're a candidate for a cash-flow crisis that may be fatal to the business. If cash flow goes up and down like a roller coaster, you may have difficulty surviving, even when volume is climbing and the work is profitable. Most builders would be better off with a smaller but more dependable cash flow.

The best strategy is to develop a stable cash flow on work that's consistently profitable. Then gradually increase the work load that's producing that profit, if possible. If not, recognize this: A small cash flow isn't necessarily bad. Big isn't always better. But a big fluctuating cash flow is always tougher to handle and probably is less profitable than a smaller but stable amount.

Of course, the right reaction to reduced cash flow is to reduce expenses. But that isn't always easy. Material and labor costs tend to vary in proportion to income. But overhead doesn't. And loan payments may not vary at all, even if business falls off to little or nothing.

One expense that can be cut quickly is your salary. It's never pleasant, but taking less for your services as manager may help your business survive until prosperity returns. When the time comes to cut overhead, look seriously at your living expenses. I think you'll find that major savings are possible.

The Tax Man

Many general contractors never make it to bankruptcy court. Instead, they're shut down by the IRS. You can't ignore the IRS the way you can other creditors. One of the quickest ways to put your company out of business is to omit making tax deposits for withheld taxes. But there are other ways to get into tax trouble, too.

There's a penalty for failing to file a return, so be sure to file. But it isn't a crime to fail to pay the amount due. There is a penalty, of course. On your personal return the IRS will charge 5% a month to a maximum of 25%. Some lenders charge nearly that much. And they want a credit check and collateral for the loan. The IRS demands neither on unpaid taxes.

If you're trying to gain time, the IRS bureaucracy is on your side. Unless your problem is assigned to a field agent for personal attention, you can get lost in the IRS computer system for at least a year without any trouble.

Let's say that you've reported a tax obligation and that the IRS has sent you a demand for payment. Here's a rule of thumb to follow. If the notice is signed by a computer, toss it out or put it in a drawer. Nothing serious is going to happen until your file's been assigned to an agent in a local field office. When that happens, you'll know it because the notice will be signed by hand.

There's only one scheme I've ever heard of that might beat the IRS out of their money. I don't recommend it. But here it is for your reference. Take every dollar out of banks, savings and loan associations and credit unions. Sell your home and all real estate. Shut down your construction business. Don't work for any employer that reports your income to the IRS. Convert all your assets into something like copper wire that can be put into storage until the IRS matter is settled. Although you won't be drawing interest on money in the bank, the wire will probably appreciate as quickly as inflation erodes the value of the dollar.

The Utility Companies

It's hard to run any business without electricity and a phone. Eventually, you'll have to pay your utility bills. But it's easy to postpone that date far longer than you might expect. First, use all of the time they allow before service is to be terminated. When that time has run out, phone the utility company. Most will postpone the cutoff date by two weeks if you promise payment by that date. If you can't get two weeks, try for ten days or one week. When that time has passed, call again and ask for a two- or three-day extension. You should have no trouble getting at least one more 48-hour extension. When the "shut-off" man arrives, write him a check for the minimum amount needed to continue service, even if the check isn't good.

Paying the bill at a local merchant that has been designated to accept payments will give you a few additional days. Utility company-approved payment locations usually take three to five days to get the paperwork and receipts to the utility office. The utility will consider your payment made at that office even if they haven't seen your check yet. Your check will take about ten days to bounce twice through your banking system. The utility will then ask you to make the check good. Plan to be unavailable for two or three days. When you finally talk to them, begin again with a series of requests for two-week, one-week, 48-hour and finally 24-hour extensions. These requests should be honored because you're dealing with a different department this time.

Eventually you have to come up with the cash. But using the method described above should buy at least six weeks of delay. That's like a six-week loan at zero interest.

Incidentally, I've found that the telephone company won't give you as much time as the utility company. Phone companies make the disconnect at their office. They don't have to send out a service truck.

If necessary, you can use this procedure every ninety days or so without adverse effect on your credit rating. If you have to do it every month, your credibility should hit bottom after about three months. That makes it tougher each month to get an extension.

Loan Kiting

Loan kiting is like check kiting, except that it's done with unsecured bank loans. You take out small loans at several banks, paying off the first loan with the proceeds from the second loan plus interest. The loans are slightly larger each time and each is repaid with interest. The idea is to build your unsecured borrowing capacity to several thousand dollars.

Loan kiting isn't illegal, provided you ultimately repay your loans. But it's more expensive than check kiting because of the fees due on each loan. The idea is to buy time until you can recover. For that purpose, it works well.

Summary

Borrowing doesn't pay off debts. It only buys you time — at a price. If you're borrowing to buy time, you'll need a current financial statement, a profit and loss statement, and your tax returns from the previous two years.

You can buy time with credit cards — which are actually prearranged loans. Stalling, floating, and check and loan kiting are other techniques you can use to buy time if you've run out of options, though they're hardly ideal and check kiting is illegal.

It's in your best interest to develop a good working relationship with your banker. During good times, borrow money and pay it back. Make sure credit will be there when bad times come around.

Dependability and profitability are the keys to managing cash flow. Big cash flow isn't necessarily better, especially if it involves extreme fluctuations. Settle for a more modest but also more dependable cash flow.

Got That Sinking Feeling?

This chapter is going to explain the bankruptcy procedure. But once I've explained it, I'll suggest other ways to get back on your feet without going through the trauma of bankruptcy.

Bankruptcy and Legal Process

If you can't find any other way out, the bankruptcy court will sort out your obligations and distribute what's left. If there's no possible way that your income could ever pay off your debts, then the courts forgive your debts, with only a few exceptions. But beware. There are penalties that come with that forgiveness.

Bankruptcy is never an easy decision. It's a last resort. Don't think of it as a way to solve all your financial problems. It's not. Here's why: You lose control of all assets beyond the minimums established by law. True, bankruptcy laws have been made more favorable to the debtor in the last five years. But it's no bed of roses. Don't believe anyone who claims different.

If you're deeply in debt and considering bankruptcy, here are some guidelines that should help you decide whether to file or not. The choice isn't as simple as you might expect. Since it's a very personal and highly emotional issue, your self-respect may be in question here. You may be a candidate for bankruptcy if you can answer these questions in the affirmative:

1) The first consideration must always be your mental and physical well-being. Are your debt problems making you genuinely ill — not just upset?

2) Are the bulk of your debt problems (at least 75%) recession-caused and therefore beyond your direct control?

3) Are the debts you're facing so overwhelming that they couldn't be resolved in three to five years?

4) Are you indebted due to the actions of others beyond your control? Partners, for example?

5) Are you indebted because of legal judgements which you lost and which can't be satisfied in three to five years?

6) Are your debt problems sucking up the assets of others? Assets you didn't accumulate? A new wife's assets from a previous marriage or an inheritance, for example?

Let's say you have $10,000 in debts that could be discharged in bankruptcy. That's probably not enough to make filing worth the penalty you pay. But other circumstances might tip the balance toward filing.

Suppose you owe the same $10,000 but have been in an automobile accident and are seriously disabled. You're a better candidate for bankruptcy if your disability will last several years. But if you have continuing cash flow from your construction company and some equity in your assets, even a debt of $100,000 might not make bankruptcy a good choice.

Use bankruptcy to discharge debts that would take the rest of your life to pay off. Don't expect to use it for personal gain or to maintain extravagant spending habits. Bankruptcy will change your lifestyle. There's no way around that.

Bankruptcy will do this for you: It will keep a roof over your head, leave you with a living income, save your automobile, household goods, working tools and $1,000 in a savings account. Nearly everything else has to go, unless it's fully mortgaged and there's no equity. Then you can usually keep it.

There are two main forms of bankruptcy proceedings. I'll explain both. First, there's Chapter Thirteen bankruptcy. It puts your creditors in a holding pattern while you work out some sort of repayment plan. But this isn't just a way of stalling your creditors. Chapter Thirteen bankruptcy is court-supervised and requires the general agreement of most of your unsecured creditors. Secured creditors, those who have a claim on a particular asset (like through a recorded mortgage), are handled on a case-by-case basis. You have the choice of giving them the security or paying off the debt and keeping the security yourself.

Chapter Thirteen bankruptcy will buy time. But the maximum time under the law is typically three years. The courts won't let you use Chapter Thirteen unless there's a good chance that all outstanding debts can be paid off within that period.

If the income available won't pay off your debts in three years, it's best to seek protection under Chapter Eleven of the Bankruptcy Act. That's called "straight bankruptcy." Creditors don't have to agree with the plan and there are no payment schedules to meet.

The final act of the bankruptcy court is a discharge of all unpaid debts. But that discharge doesn't happen automatically. You have to petition the court for relief and convince the court that discharge is proper. The court has the choice of granting or not granting your petition.

Bankruptcy procedure is complex: don't expect to handle it without a lawyer. Some books say you can, but I wouldn't advise it. There's more to it than meets the eye. Find an attorney who handles a volume of bankruptcy matters and watch him carefully to make sure he performs. Attorneys are busy, just like everybody else. Bankruptcy attorneys sometimes have more low-profit work than they can handle. If you expect to move quickly through the maze of legal requirements, be the "squeaking wheel" that gets the grease.

Here's what to expect if you file for Chapter Eleven bankruptcy. First, you fill out a detailed bankruptcy petition. This shows all debts that you claim can't be paid. Attach to the petition a list of all assets that can be used legally to satisfy those debts. You don't have to list any debt that you plan to pay outside of bankruptcy. But once you've filed your petition with the court, it's too late to add a debt to your list. You are prohibited from filing for bankruptcy for the next seven years. So any debts not discharged in the first bankruptcy action become continuing obligations.

The next step is to file the petition with the court. From the date of that filing you're officially bankrupt. Some people believe that bankruptcy begins from the date of your first court appearance. That isn't so. From the moment you file the petition, all suits against you, as well as collections and foreclosures, are suspended. Creditors have to wait until the court sorts things out in its own good time.

Once the petition is filed, the court will send out a notice of hearing to all creditors on your list. When the hearing date arrives, the court selects a trustee to handle the bankruptcy proceedings. This is known as the "first meeting of creditors."

The trustee liquidates your assets as permitted under the law, and distributes the proceeds according to established priorities. This list sets down the order in which debts have to be discharged. The tax collector gets his money first, as you might expect. State, county, and city governments, landlords, and employees come next. Once all these parties have been satisfied, your other creditors pick the bones of what's left.

Right up there with the IRS on the priority list are the court trustee, the court and the attorneys. You can see why creditors like your material suppliers and the phone company don't do very well in bankruptcy court.

The first meeting will probably be your last bankruptcy session. The rest are handled by the trustee of the court. Expect your bankruptcy proceeding to last from six to nine months.

Once you've filed the initial petition, you can buy and sell real estate other than assets involved in the bankruptcy. You can borrow, enter into contracts, and operate normally again. That's the good news. The bad news is the burden you carry as a bankrupt individual.

There is usually loss of self-respect and reputation. By business standards, you have failed. You aren't up to par. You've been tested and found wanting. Your ex-creditors trusted you and now they have to suffer because of that trust.

All that may be true. But it doesn't help you become a productive and responsible member of the construction community again. After all, we're all contractors-in-training. Sometimes we make mistakes. That's inevitable. It isn't the mistakes we make, but what we learn from those mistakes, that's important. Some of today's most successful contractors have gone through bankruptcy. They learned from their mistakes and used the experience to improve the way they do business. You can do the same.

Look at it this way. If you go under, you're going in good company. Both New York City and Chrysler Corporation might have ended up there if the government hadn't lent a hand.

Another problem to cope with is your credit rating. Bankruptcy is going to follow you around for several years to come. You have to re-establish good credit all over again. It's like you're a teenager again without credit cards and charge accounts. But it won't be as hard to get set up this time because you've been through it all before. At least you understand how to do it.

But don't expect the fact of your bankruptcy to disappear as time passes. Every time you apply for credit, you have to reveal the bankruptcy. The record is passed on from creditor to creditor. And there's nothing you can do to erase your bankruptcy from the record of credit-reporting agencies for several years. The best you can do is stay current on all obligations in the future.

Suits, Judgements, Liens, Compromising Debt

Legal entanglements are frustrating, awkward, time-consuming and expensive. Stay away from lawyers if you can. But sometimes there's no practical alternative to filing or defending a lawsuit.

A contractor fighting for financial survival has a major disadvantage in court. His financial condition is completely irrelevant. The court hasn't the slightest interest in the condition of the economy or the condition of your finances. That's your problem, not the court's.

The fact that a lawsuit might tip you over financially is also irrelevant. All that concerns the court in a lawsuit is the debt itself and the facts that created that debt. Do you owe it or don't you?

But courts also take a narrow view on some other issues. And this time it's to your advantage. There's a principle in the law that no one can sue for losses that have not actually occurred yet. Here's an example: Suppose you contract out some work to a subcontractor. You get paid for the work but can't pay the subcontractor. The sub sues the owner to collect for the work done and gets a judgement for the full amount. The owner refuses to pay on that judgement and sues you because he's already paid you. The law says there is no ground for the owner's suit yet. The owner has to pay the sub before he is deemed to have suffered a loss. Only then can he sue you to recover for the double payment. The judgement itself isn't a payment. It's only an obligation.

This doesn't mean that the owner can't turn you in to the Contractor's Licensing Board. It only means that a legal right to sue does not exist until an obligation has been paid — until a loss has really occurred.

Here's a chronological rundown on what happens in a lawsuit. Actual practice varies in some states. But generally the procedure will be as follows. (The major exception is small claims court. In small claims court, the procedure is simplified in many respects.)

The first step is service of process. You'll be handed a packet of papers that have been prepared by the attorney for the plaintiff (the person who has the complaint against you). Among these papers will be a *summons* and *complaint*. The summons tells you how much time you have to file an answer to the complaint and where that answer has to be filed. The complaint explains in legal language exactly what the plaintiff is complaining about and what he wants the court to do.

Make a copy of every page of the summons and complaint for your attorney. Write on the summons when you were served. At this point you have three options: First, you could reach a settlement with the plaintiff within 30 days of service. Second, you could file an answer to the complaint to indicate that you will contest what the plaintiff claims. Third, you could do nothing. Making no response gives the plaintiff the right to take a default judgement. That's the legal equivalent to admitting that you have no defense to the complaint. If you accept the default, expect to see the sheriff try to pick up some of your assets within about 60 days.

If you file an answer to the suit, you set in motion a chain of events that must occur before there's any obligation to pay the amount requested. You'll probably want to have an attorney prepare an answer for you. If so, forward a copy of the summons and complaint to that attorney right away. Remember that there's a 30-day time limit on filing the answer.

It's perfectly acceptable to send copies to several attorneys. Have each suggest the correct action to take and quote a fee for doing the work required. Then you select the attorney that seems to have the best plan and the most reasonable fees.

To confuse matters, consider filing a countersuit against the plaintiff. This makes your plaintiff file an answer and puts some risk into the suit from his standpoint. This is called a *cross complaint*. It makes resolving the suit more difficult, but it also increases your legal fees.

Once your creditor has sued, you've answered, you've countersued, and he's answered, it's time to prepare for trial. This process is called *discovery*. It usually involves depositions (taking testimony under oath at the office of one of the attorneys) and interrogatories (written questions exchanged by the parties). When this process is completed, the plaintiff will file an *At Issue Memorandum* to indicate to the clerk of the court that the parties are ready for trial — that is, they have finally come to the issue.

The clerk assigns the first available court to try the case. Usually that will be months away. See Figure 4-1 for a scale depicting the amount of time consumed by various court procedures.

When your court date rolls around, you may decide to go through with the trial, or to attempt to settle without trial. The courts encourage settlement and may pressure the parties to settle without trying the case. At least 95% of all cases filed are settled without trial.

If you lose at trial, the plaintiff gets a judgement against you. That's like a hunting license. He still has to find some of your assets and get the sheriff to seize them on his behalf. If he can't find assets, he can require you to attend a debtor's examination and answer questions under oath on what you own and where it's located. Most states have a limit on how often these examinations can be scheduled. Usually it's every six months. In between examinations, they have to leave you alone.

The sheriff does the actual work of seizing assets. He can *levy execution* against bank accounts or personal property such as a car, furnishings and equipment; or garnish wages in the hands of an employer. But this is expensive. The plaintiff has to post a bond of several hundred dollars before the sheriff does anything. Some of these costs can be recovered against the judgement debtor when the assets are sold. Generally, the larger the asset to be seized, the more expensive it is for your creditor to seize it. But that doesn't mean he won't try.

The law in many states permits installation of a keeper at the debtor's place of business. A keeper is a uniformed sheriff's deputy. He goes to the debtor's place of business and can take any of several actions, depending on the creditor's instructions. The keeper can make a till tap — pick up any cash or negotiable instruments belonging to the business. He can also take possession of the premises for the day, seizing any cash that comes in during the day. Finally, the keeper can take possession of the entire premises for 24 hours. At the end of that time, the keeper calls for bids from a moving company to move all the debtor's furnishings and equipment into storage. The creditor has to post a bond to cover costs of moving and storage. That may be several thousand dollars for a going business.

The keeper can be authorized to accept a certain amount of cash from the debtor to satisfy the debt. And that's usually what happens. Nothing disrupts a business faster than having the sheriff unroll his sleeping bag at the reception desk and announce that he's taking possession of the premises. That brings a debtor to his knees in a hurry.

Keepers are expensive. In a larger business one keeper is needed for each exit. The minimum

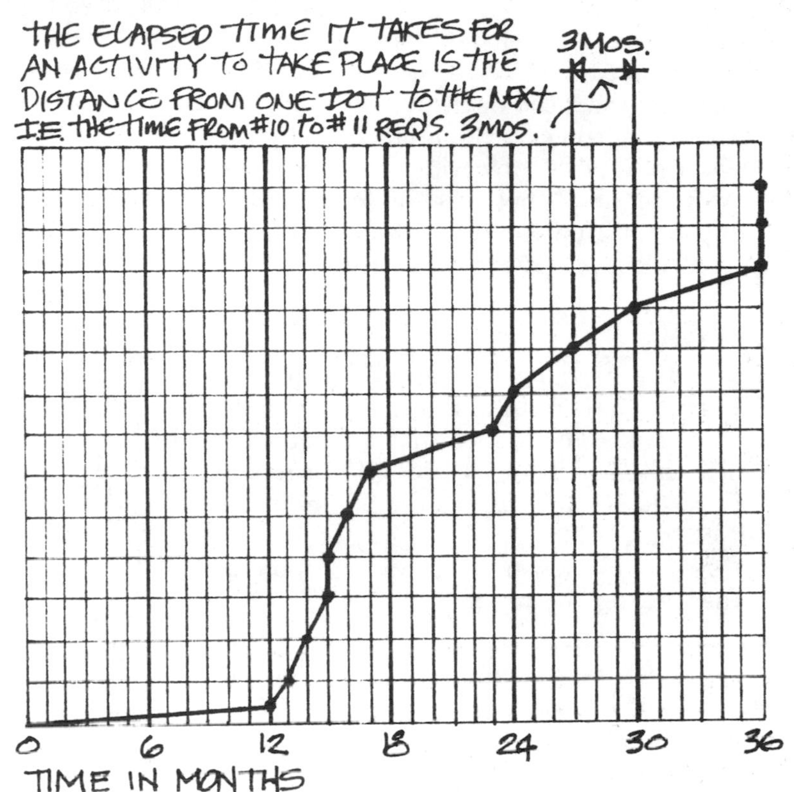

Lawsuit time consumption
Figure 4-1

deposit is usually at least $500. But generally, it takes only one day to collect the debt. The unused balance will be refunded.

It isn't cheap to find and seize assets. If the debtor has none available, or if the creditor doesn't have a deep pocket, it may be wiser to just record an abstract of judgement with the clerk in your county. If the judgement debtor ever tries to sell land in that county, the abstract will show up in the title search as a lien on the property. The buyer will then insist that the lien be paid off before he takes title.

If you have few assets and little income, your creditor's efforts may be wasted. He can't get what you don't have. If you lease your cars and trucks and homestead your house, you leave very little that is accessible to a creditor. Converting vulnerable assets into cash, a commodity, or selling them to a friend may discourage your creditor to the point that he gives up the chase.

But even if the creditor gives up, the judgement is still valid and earning interest at the rate established by law. Once you're back on your feet and making money, the creditor can come after you again.

Let's say that you elect to settle rather than go to trial. You and your attorney agree on a settlement amount with the plaintiff and his attorney. The plaintiff will probably want that settlement written up as a *consent to judgement* with a payment schedule. A consent to judgement is more effective than a simple promise to pay. To enforce the latter, the plaintiff would have to sue all over again. A judgement, on the other hand, can be entered directly and enforced by levy without a trial.

As long as you meet the payment schedule, the judgement never becomes a public record. If you default in your payments, however, your creditor's attorney will request that the judgement be entered. But it's still up to the creditor to collect. And he has to locate assets, just as before.

Mechanic's Liens

Mechanic's liens are strange animals. The way they work varies from state to state. But the broad principles are the same. The idea is that anyone who has furnished labor or materials to build or repair something should be able to use the value in the thing built or repaired as security if he doesn't get paid. That's a good idea. But it's the application of that idea that creates the problem.

A lien is a security interest, like a mortgage. But unlike a mortgage, mechanic's liens don't have a power of sale. The lienholder can't sell the property. He has to be satisfied with holding the lien until the owner wants to clear his rights to the property.

Liens give the lienholder a right against the owner — the right to get paid before title to the property is clear. In practice, the lienholder gets paid when the property's sold to a new buyer or when the owner wants to take out a new loan on the property. Otherwise the seller can't give clear title to the property, and the new lender would be subordinate to the lienholder.

Some states have special requirements that must be followed before lien rights can be enforced at all. For example, anyone without a direct contract with the owner may have to present a *Preliminary Notice of Intent to Lien* to the owner, the general contractor, and the lender. If you don't do this within 20 days of starting work, you may lose all lien rights.

Here's the biggest problem with mechanic's liens: The system works only if everyone lies and then uses the money those lies buy to pay exactly who should get paid. The subs don't get paid until they give the general contractor a labor and material lien release. But the subs can't get releases from suppliers and tradesmen until they're paid. But if the sub hasn't been paid, how can he pay anyone else? He can't. So each sub must trust you and sign a release before actually getting paid. Each subcontractor has to lie to the general contractor by saying that he's already paid all his labor and material bills. That's the dilemma of every contractor who's running a business on limited capital. There's no easy way around this problem.

If the general contractor gets paid but doesn't pay his tradesmen and material suppliers, the tradesmen and suppliers can lien the owner's project and sue the owner for payment a second time. If they've filed the Preliminary Notice within the required twenty days, they'll probably win.

The best way to avoid problems with a mechanic's lien is to have enough money to pay all bills before getting paid for the work done. But not many small contractors can do that.

Compromising Debts

Compromise is usually the best way to settle disputes. That's why compromising debts is a good way to handle bills that can't be paid. If you can't pay the whole bill, then why not pay half and call it settled? Believe me, it's done all the time. Actually, that's all that happens in a bankruptcy. The court-appointed receiver liquidates your assets and distributes the proceeds as far as they'll go to all creditors. Most general creditors will get only a few cents on the dollar.

Offer to compromise with your most demanding creditors. Be as persuasive as possible. But be sure that any payment you make is in full satisfaction of the whole amount owed. You don't want them to claim later that the money received was only "on account," with the remainder to be received later. Note on your check that endorsement acknowledges compromise of the debt and that the amount paid is accepted as "payment in full" on a particular job or invoice.

Unfortunately, some real estate developers make it a habit to pay fifty cents on the dollar. They contract with a builder to do a job and then delay payment because of some real or imagined discrepancy. When the contractor is buried financially and strapped for cash, they offer him fifty cents for each dollar owed. The offer usually comes with a challenge: "Sue me and I'll bankrupt on you."

The contractor has very little choice. He's out of money and can't afford the legal fees needed to chase the developer. Compromise may be the only way to keep operating. It's a matter of survival.

Notice that there's a basic difference between compromising your debts because they can't be paid in full, and planning from the beginning to pay less than the contract amount. Here's what I do when I discover a "fifty-cents-on-the-dollar" operator. First, I decide that I've made a mistake in working with this guy and resolve never to do it in the future. Next, I decide to do without the money for the time being. Then (because my attorney is understanding when it comes to getting paid) I go after this guy every way possible. That includes putting liens on his clients and his projects, attaching his wages — and phoning him every day, twice a day, even on weekends and holidays. This repetitive phoning is a very effective tool. I've had a developer pay me off just to get me off the phone!

Asset Protection, Trusts and Corporations

Let's say that you're considering bankruptcy. Or maybe you're involved in a big lawsuit that you can't win. Either way, you stand to lose whatever assets you've built up over the years.

There are some basic points to understand about protecting assets. First, if you've transferred assets to friends, relatives, or business associates, the courts can view this as an effort to defraud your creditors. The court has the power to freeze assets in the hands of your friends, associates, or relatives; and can issue a deed to property if necessary. This could prove very embarrassing to your friends and expensive to you. But there are acceptable ways to transfer assets.

If you're going to transfer assets to protect them from creditors, do it long before you come to trial. Make each transfer as much as possible an ordinary business transaction. Just hiding assets won't deceive anyone.

And, of course, holding the proceeds of liquidation in a bank account is inviting court action. It can also create tax problems.

One way to protect assets is to put them into an irrevocable trust with someone else as trustee and beneficiary of the trust. The court regards the trustee as an uninvolved third party. The difficulty with this arrangement is that you lose control over what happens to your assets. Obviously, you can't sell them. And it's next to impossible to borrow against assets held by a trustee.

There are several types of trusts: the testamentary trust, the revocable living trust, the irrevocable trust and the pot trust. There are many variations, but these are the basics. Incidentally, only the irrevocable trust can help you protect assets. The other trusts are accessible in one way or another. As a rule, any trust that gives *you* the power to remove assets is also accessible to your creditors. Get specific information from the trust department of your local bank, or from an attorney who handles estates and trusts.

Trusts have been used for many years to preserve assets for future generations. The problem with trusts is that they don't protect your assets fully until you're dead.

A third alternative is to sell all assets and convert the cash into a commodity that can't be traced, doesn't require upkeep, and isn't affected by economic cycles. Food, for example, is unsuitable because it deteriorates. Stocks and bonds won't do because ownership is recorded. Equipment might be a good bet, but in time it may become obsolete. A product in its rawest possible refined condition is a good bet. Pure copper in ingot form fits this description. It won't deteriorate, requires little care, and ownership isn't recorded. It's too heavy to carry off in a bag, and may increase in value as the dollar inflates.

If you convert assets into a commodity like copper ingot, keep it out of state, out of sight, away from relatives, and under lock and key. The less you say about the commodity, the more secure it will be. I would also suggest that you let considerable time pass before converting it back to cash.

The corporate form of organization can help you protect assets. In fact, the corporation was originally developed to provide limited liability for investors. Each investor is liable only up to the amount of his investment. More than that, the corporation's creditors can't reach.

The corporate form offers some protection from lawsuits. A big judgement against the corporation may send it into bankruptcy. But you walk away free of obligation even if you were 100% owner. It's the corporation that's responsible, not you personally.

There's danger here, however. Nothing in the law is that simple. First, a corporation is only valid if it's formed properly, operates as a separate legal entity, and has adequate capital. The problem is that no one can really define what constitutes an adequate amount of capital. In court, the plaintiffs may "pierce the corporate veil" by showing that the corporation was only a paper sham that had no real existence of its own. If they manage to do that, they've demolished all protection the corporate form might offer.

Another difficulty is that most suppliers will want you to sign personally for all major purchases and loans. No one will accept the corporation's credit until the corporation has considerable assets. Also, note that anyone suing the corporation can name you personally as a second defendant. That's a perfectly acceptable practice.

This leaves you with the problem of showing, in court, that it's the corporation that's responsible and not you. You have to prove it and pay for the proof. It's not automatic.

Finally, the corporation has to pay income taxes, just as individuals do. Usually there's a minimum tax of several hundred dollars. Also, you'll need to hold at least one annual stockholders meeting and keep corporate minutes. All of this takes time and money.

You may decide that running your business as a corporation is a nuisance. Unless you have a company bigger than mine, forming a corporation may be more trouble than it's worth — at least in the context of protecting assets.

If you do incorporate, it's critical to separate your assets from the assets of the business. Don't mix corporate and personal funds. Have an invoice or payroll record on file for every check the corporation writes. Don't use corporate assets for personal purposes. Make sure every dollar you take out of the corporation is recorded as either salary, reimbursement, or dividend.

One last note before we leave the subject of corporations. If you're having trouble making payments on a piece of land and you're on the verge of losing it, consider this: form a corporation and transfer the property to the corporation. After a while, put the corporation in Chapter Thirteen bankruptcy.

The court will put the corporation's creditors on ice for a few months while you devise a payment plan. Under Chapter Thirteen the court offers protection for three years while you pay off creditors. Creditors have to conform to your ability to pay. This may reduce land payments substantially and give you several months to come up with a payment plan. It could be time enough to avoid losing the land's equity. With any luck, you'll sell it at a profit long before the three years have passed.

Homesteads and Foreclosures

Many builders use the Homestead law to protect their primary residence from attack by creditors. The amount of equity protection the law provides has been boosted upward several times in recent years. Originally, the homestead limit was set at $7,500. It was revised upward to $12,500, then $15,000, $20,000, $30,000, and finally $40,000. These changes reflect increasing home values. Today, a homestead will protect the first $40,000 in equity.

It's easy to declare a homestead. Your local stationery store probably has a standard Declaration of Homestead form for your state. Fill in the appropriate blanks for address, assessor's parcel number, legal description, name, and so on. This information will be on your tax bill. Have your signature notarized and then take the document to the County Recorder's Office for recording. *Congratulations*. You've just protected the first $40,000 of equity in your home. That's all there is to it.

If the equity you have is greater than $40,000, the balance is subject to claim by your creditors. But there's another maneuver to protect that.

The most practical method of protecting the balance is to encumber it somehow. You can simply borrow all equity out of the house. If you can get a high appraisal, pull the full equity out, except for the protected $40,000. Look at Figure 4-2. In this example, you can protect your equity by refinancing the first trust deed and taking out a second for the balance of the equity.

The problem with this maneuver is that you have to pay the loan fees up front, make monthly payments on the loan, and you may have to hide the loan proceeds to keep them away from your creditors.

An easier solution is to keep your present mortgage and get a friend or relative to record a large second mortgage. This doesn't have to involve any actual cash transaction. Just record the mortgage on the property, showing that all of your equity, except the homesteaded $40,000, has been exhausted.

In Figure 4-3, we begin with the same situation as in Figure 4-2. But instead of refinancing the first trust deed, you can create a new second trust deed in the amount of $40,000. This can be a loan from a bank, a friend, or a relative. Now you have only $22,000 in unprotected equity. You can cover that with a new third trust deed. You probably can't get

REFINANCED EQUITY PROTECTION

EXISTING CONDITION:	HOME VALUE	$175,000
	FIRST TRUST DEED	73,000
	EQUITY UNPROTECTED	102,000
	HOMESTEAD PROTECTION	40,000
	UNPROTECTED EQUITY	62,000
REFINANCED CONDITION:	HOME VALUE	175,000
	75% FIRST TRUST DEED	131,000
	UNPROTECTED EQUITY	44,000
	HOMESTEAD PROTECTION	40,000
	UNPROTECTED EQUITY	4,000
	NEW SECOND TRUST DEED	4,000
	REMAINING EQUITY	$ 0.00

Refinanced equity protection
Figure 4-2

that from a bank, so you may have to depend on a relative. If you've already borrowed from your family, record that debt as a third trust deed. It gives them some collateral and helps protect your equity. By doing all this, you leave your first mortgage intact and unaffected; and your equity appears non-existent and can't be liquidated in bankruptcy.

A new wrinkle to appear on the scene is the Automatic Acknowledgement Statute. This means that you don't need to record your homestead document. The court recognizes your homestead rights automatically, without any effort on your part. To be on the safe side, though, I'd fill out the homestead form and record it. The security of your home is too valuable to take lightly. Do all you can to protect it.

A financially-troubled contractor with property to lose should know everything possible about foreclosure proceedings. He's going to need it. Unfortunately, the exact details are different in every state. I'll outline the broad concepts here as they

ADDITIONAL T.D. EQUITY PROTECTION

EXISTING CONDITION: SAME AS FIGURE 4-2

NEW CONDITION:

HOME VALUE	$175,000
FIRST TRUST DEED	73,000
UNPROTECTED EQUITY	102,000
HOMESTEAD PROTECTION	40,000
UNPROTECTED EQUITY	62,000
NEW SECOND TRUST DEED	40,000
NEW THIRD TRUST DEED	22,000
REMAINING EQUITY	$ 0.00

Additional trust deed equity protection
Figure 4-3

apply in California, where I do business. Get specific guidance from a lawyer familiar with foreclosures in your state.

There are two kinds of foreclosures: judicial and real estate. We'll concern ourselves only with the real estate foreclosure.

Real estate foreclosures are used with real property that has been secured with a deed of trust. There's a misconception that foreclosure proceedings can't be started on a piece of property unless you're at least ninety days behind in your payments. That's not true. Here's the way it really works.

If your property payment is due on the fifth of the month and you don't make it, legally you're in default. On the sixth of the month, your mortgage holder can serve you with a Notice of Intent to Foreclose. Once he's done that, the clock is running. You now have ninety days in which to make up those back payments, plus any trustee's fees and recording costs, or taxes and assessments if applicable.

Normally, a mortgage holder won't send you a foreclosure notice until you're at least ninety days behind in your payments. But it's entirely up to him. He can be lenient or not, as he chooses. So, be nice to your mortgage holder. People don't like to foreclose on friends. If, within the ninety-day redemption period, you make your payments, that's the end of the story. If, on the other hand, you don't pay, the property is ordered sold at auction. The date of the sheriff's auction is typically thirty days after the order is issued by the court. Up until the actual time of the sale, or a couple of days before, you can still redeem your property by bringing payments current and paying the related expenses.

After this time, you must come up with the entire amount of the loan and related costs in order to save your property. You'll need a new loan, and this is difficult at best, especially when you're going through foreclosure. There's danger in saving a drowning man. The victim may drag you under, too.

Summary

Bankruptcy is never an easy decision. Use it only if you have no other alternative. Bankruptcy offers protection from creditors; and the prospect that debts will be adjusted or forgiven completely. But there are penalties: loss of self-esteem and a bad credit history.

The threat of bankruptcy is a powerful weapon in keeping a creditor from suing you. He doesn't know what financial shape you're actually in. Lawsuits, however, can actually work in your favor if time is what you need. To gain more time, file a cross-complaint. If the court decides against you, remember that seizing assets is expensive to the creditor. Many creditors will only record the judgement, hoping to get paid eventually.

If you intend to liquidate your assets, do it long before trial. Consider converting the cash into a commodity that can be stored out of sight. Trusts, corporations, and homesteading are other alternatives for protecting assets. The purpose of this chapter has been to show you how much time you can buy before going through bankruptcy. If you can buy enough time, you can probably work your way out of debt. I offer this advice to help you get back into financial health and pay your creditors, not to beat them out of their rightful earnings. They work hard for their money, just like you do.

5 Who's Minding the Store?

Someone has to run a construction business if it's going to survive. By that, I mean that someone has to take care of office chores to support work that's done in the field. In construction contracting, paperwork and recordkeeping are as important as concrete and lumber. A builder who doesn't understand that would probably be better off back at the construction site as a tradesman or foreman rather than as a contractor.

This chapter will focus on paperwork and recordkeeping: accounting, journals, registers, withholding, billing, insurance, and bonding. We'll also spend some time on how to detect employee theft. My emphasis will be on doing just an adequate job, with the least effort, and in the shortest possible time. You don't need great records. Adequacy is sufficient. And you don't want to spend one second, or one dollar, more than necessary on paperwork. But no successful construction company can survive for long with inadequate records.

Accounting

Unless you're an accountant, accounting is no fun. In fact, accounting ranks alongside of a root-canal job at the dentist on my list of work to avoid. Unfortunately, it comes with the territory. If you don't like accounting or can't at least learn to tolerate it, you'll always have trouble running a construction company.

Few people ever went bankrupt with good records. Accounting helps you anticipate problems and avoid trouble before it arrives. That's what a balance sheet and expense statement do for you.

If you aren't good with numbers, have your spouse keep the books or hire a bookkeeper. If you don't want to supervise a bookkeeper, have an accountant maintain your ledgers. In any case, you'll probably need an accountant at year-end to prepare the company tax returns.

It's more important that your books be kept up to date than that they meet every conceivable accounting standard. Business success depends on knowing exactly where you are financially at any moment. If your books and ledgers are always in your accountant's office, you may find yourself waiting in line to check out your balance. Keep your accounting ledgers in your office. Make the daily entries yourself, if necessary. Your accountant needs them only for quarterly reports, periodic filings, and year-end tax returns.

Whatever you do, don't put it off. That's the worst thing possible. If you let the paperwork go for thirty, sixty, or ninety days, you won't remember where the money came from and how it

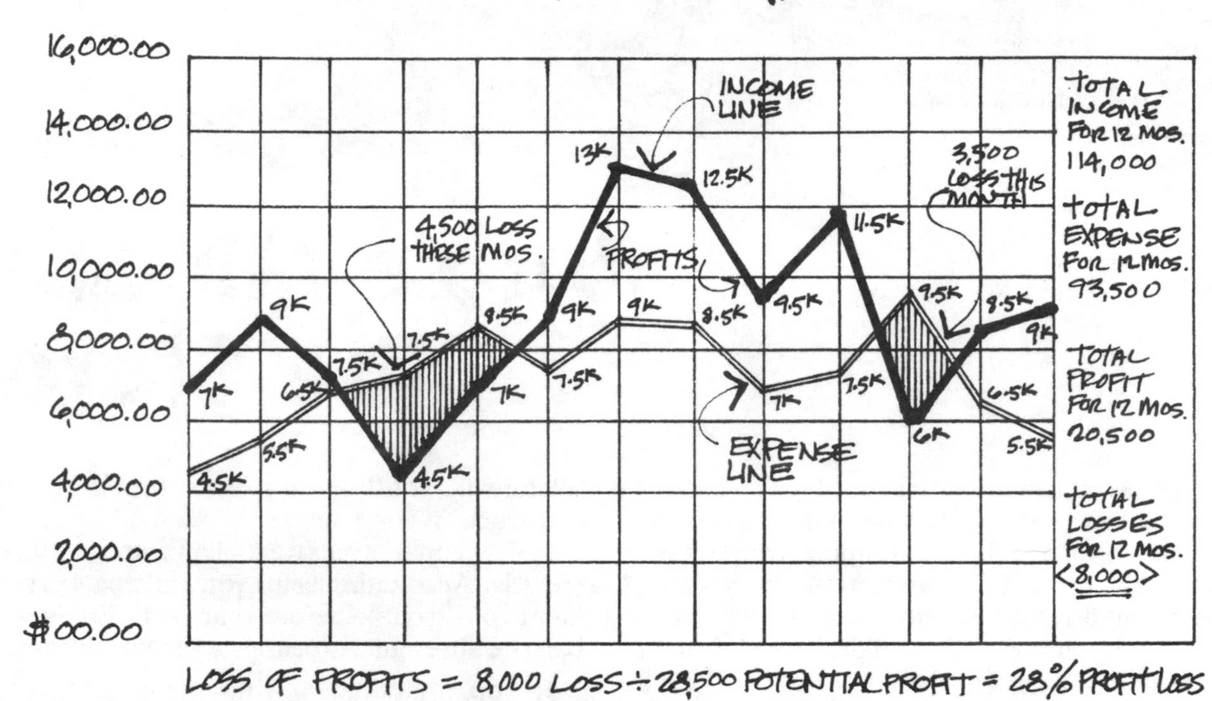

Income vs expense record
Figure 5-1

was spent. Any income that you can't show was spent for a legitimate business purpose is taxable income to either you or your company. *Stay current on your bookkeeping unless you enjoy lengthy discussions with an IRS agent.*

Fortunately, doing your own bookkeeping in a small construction company isn't an impossible task. Most of it is just being able to add, subtract, multiply, divide, and figure decimals and fractions.

Do your accounting the easiest possible way: you don't need elaborate journals. Avoid using a computer, if possible. Keep it simple. What you need is records adequate to meet IRS rules: a check register, a monthly total on expenses and income by category (a profit and loss statement), a balance sheet (showing assets and liabilities), and a listing of activity by account number.

One additional record you'll find very useful is shown in Figure 5-1. It's a chart that plots income against expenses. Update it monthly. The chart helps you understand your financial condition at a glance. You know what kind of deals are needed to keep you profitable. Update this graph monthly so you know when you're apt to take a loss. That helps you anticipate borrowing needs. If you're in the loss column, you can predict how long you'll be in the red and how much profit is needed to get back in the black.

Your Checkbook

Every business needs a checking account. Once you

have one, there are only four things you can do with it: You can put money in it. You can write checks and take money out of it. You can reconcile it at the end of each statement period. And, finally, you can get it hopelessly fouled up.

Reconciling a checkbook is what you do to keep a checking account under control. It's just a fancy name for balancing the account — finding errors and determining the exact balance available in the account.

Reconciling a statement is really very simple. When the bank sends you a statement of activity at the end of each month, go through it to see what checks cleared that month and what checks were still outstanding during the period. That's the first step. Be sure the beginning balance on the statement is the same as the ending balance on the prior statement. See if the checks that the bank cleared were all your checks and were listed correctly. Be sure all deposits made for the month show up on the statement. Satisfy yourself that all service charges and fees are legitimate.

The final and most important step is to find the balance available in the account. Start with the statement balance. Add all deposits made but not shown on the statement. Subtract all checks written but not listed on the current or prior statements. The result is your available balance.

That's really all there is to it. But if you're still a little confused, never fear. The reconciling procedure is probably outlined on the back of the statement. If it isn't, trot down to the bank where you have your account. They'll gladly show you how to do it.

By the way, in accounting, a number such as $38.50 standing alone is assumed to be a positive number. If you see $38.50 preceded by a minus sign, that's a negative amount. Numbers in brackets such as *($38.50)* or printed in red are also minus figures. That's called being in the red. Black ink is good, red ink is bad — except on your tax return.

Cash Journals

Every business needs a cash journal to record all income. See Figure 5-2.

A cash journal is nothing more than a list of your company's earnings. You can use just about any kind of form. A cash journal makes it easy to find your total income for the month or for any period. This is a big help at tax time. I use a standard receipt book that I bought at a stationery store. Make an entry for every dollar of income received. And deposit all receipts in your company checking account. Because if you deposit part of a check and take cash for the remainder, your bank statement becomes an inaccurate record of receipts. So keep your checkbook accurate. Deposit all of your funds and write a cash check back to yourself so everything is fully traceable.

Client Record

Next, you'll need a client record. This is a listing of all invoices sent to and income received from each client. See Figure 5-3.

In the client record, enter each transaction on a separate line. For each invoice, show the invoice number (if any) and the date of the invoice. Write the amount of the invoice in the debit column. When you're paid, enter the payment and the date you received it. The amount received goes in the credit column. In the third column keep a running balance of the amount owed, adding and subtracting debits and credits as they occur. This way you always know what your clients owe at any time.

If you have to adjust an invoice that's already been sent out, make up a credit invoice and show it in your client record as if you'd actually received a payment from the client. Enter a credit invoice in the credit column of the register for that client. Don't try to change an invoice that's already been sent out.

It's a good idea to photocopy any checks your clients send you. That way you have their bank name and checking account number in case you have to sue to collect.

That's all there is to keeping track of activity for each client.

Check Register

Next comes the check register. It's probably the most complex accounting record you have to keep as a construction contractor. But don't let it scare you. It's easy too.

A check register lists each check written and why. For the check register, use ledger sheets with vertical columns. Most good stationery stores sell sheets that will be perfect. One sheet may be big enough to list all checks written in a month. See Figure 5-4.

CASH JOURNAL

DATE	SOURCE OF INCOME	AMOUNT
5.3	BOB JAMISON	$ 3367.18
5.9	A & J SUPPLY	18.55
5.9	CAMDEN COUNTY	10,087.73
5.19	MR. & MRS. HAMPTON	1,103.41
5.26	BOSTROM FINANCIAL	8,374.27
5—	TOTAL MONTHLY INCOME	$22,951.14
1.1 to 5—	TOTAL INCOME YEAR TO DATE (INCLUDE ALL PREVIOUS MO. TOTAL)	$97,881.56

Cash journal
Figure 5-2

CLIENT RECORD

CLIENT NAME: BOSTON FINANCIAL CORP.					
NO.	DATE	ITEM	DEBIT	CREDIT	BALANCE
1.	3.7	INVOICE #257	6374.88		6,374.88
2.	3.14	PMT. CHK.# 2057		5000.00	1,374.88
3.	3.14	INVOICE #281	7499.03		8,873.91
4.	3.17	CREDIT INVOICE #293		847.11	8,026.80
5.	3.22	PMT. CHK.# 2234		6,000.00	2,026.80
6.	3.31	MARCH SUB total	13,873.91	11,847.11	2,026.80

Client record
Figure 5-3

CHECK REGISTER

DATE	CHK NO.	PAYEE	CHK. AMT.	ACCTG.	LUMBER	PAY-ROLL	HARD WARE	CONC.
3·2	687	MIDLAND LUMBER	873.07		873.07			
3·2	688	JACKSON HARDWARE	37.52				37.52	
3·5	689	ELLEN JACOBS	85.00	85.00				
3·6	690	BOB HANSON	373.80			373.80		
3·6	691	MADISON CONCRETE	405.10					405.10
3·12	692	MIDLAND LUMBER	123.15		123.15			
3·15	693	MIDLAND LUMBER	11.17		11.17			
3·15	694	BAKER SUPPLY	159.22				159.22	
3·27	695	BOB HANSON	373.80			373.80		
3·29	696	LARRY BISHOP	241.11			241.11		
		MARCH SUB total	2,682.94	85.00	1,007.39	988.71	196.74	405.10
1·1 to 3·31	–	total to DATE	8,717.14	641.10	19,681.11	22,907.11	921.09	6,784.77

Check register
Figure 5-4

A check register isn't much different from a client record. In the check register you enter the date, check number, who the check was written to, and how much the check was written for.

There should be room in the check register to spread expenses among the expense categories you use. A heading at the top of each column shows what expenses are listed in that column: Headings might be *Concrete, Lumber, Payroll, Subcontract,* or use numbers like 100, 200 or 300 (if you think like an accountant). It doesn't make ten cents worth of difference in the long run. Use any system of headings that you can understand. And keep it simple.

At the end of each month, draw a line across the page below your last entry. Total all columns. In the *check amount* column, you get a total of all checks you've written that month. You could do the same thing in your checkbook if you wanted to. But the check register shows all checks at a glance and distributes expenses among the categories you select. That isn't possible in your checkbook.

The remaining columns show totals of the categories you have identified with headings: insurance, concrete, plumbing, electrical, accounting, hardware, lumber, heating, roofing, equipment, or anything else you want to segregate out.

Before we leave the check register, let me interject one word of caution. Even if you're running a sole proprietorship, don't mix business and personal funds. For tax purposes, you and your sole proprietorship are one and the same. But keep all personal transactions out of your business account and all business expenses out of your personal account. Keep them separate. That makes your accounting and tax reporting much simpler.

Here's another thing to consider. Every partnership and business you run needs a separate set of books. You can't put them all in one ledger. That would be co-mingling of funds. Bookkeeping isn't like making goulash. You can't dump all the ingredients into one pot. Instead, keep the peas, carrots, onions, and potatoes each in piles of their own.

Job Costing

Job costing is like accounting in many ways. But the account categories are different, and the scope is limited to a single job. Your job cost record should show what's been spent for each trade or part of each project. Adding up all cost categories to date will show total expenditures to date for the entire project.

There are three important reasons for keeping job cost records. First, your job cost records should be the foundation for every future estimate you make. The best estimating data you can find for your next job are the actual costs of your last job. Second, without accurate job cost records, you don't know how much has been spent on a project to date, or how much more you can spend and still make a profit. Third, job cost records are the basis for invoices you issue to clients.

The essential thing in job costing is to pick up all your costs as money is disbursed and record each expenditure to the proper category. Job costing is a constant process. Do it daily. If you're the type of builder that lets his receipts pile up, don't. You'll never know where you are financially if you don't stay on top of your paperwork.

Setting up a job costing system doesn't have to be difficult. Create classifications, or categories, for each part of the job you want to record separately. That's not much harder than setting up your checkbook. That's what a job cost record really is — a giant check book. Figure 5-5 shows an example.

The job cost record should record each expenditure by job name, date, check number and the amount of money spent. Then decide which category that check fits under: wall framing, roofing, floor sheathing, footings, or interior trim. This is known as "spreading your costs." Make the categories as broad or as detailed as you want. Unfortunately, one check could be distributed among several categories at times. In that case, either issue multiple checks or place the cost in the largest applicable category.

Once the cost is recorded correctly, you can pull a subtotal on any individual portion of the job without having to go through all of your records. You can also pull a total cost for each month by totaling the cost sheet for that job. Do this one to three days before you plan to send out invoices. Bookkeepers call this the cut-off date. Finally, you should have a running total of the costs on the job to date.

Payroll

Now that you have a fair idea of the bookkeeping and job costing required, let's look at the payroll records you need.

A contractor's largest single expense is almost

JOB COST RECORD

PROJECT NUMBER: C154 HEAVENLY CANDY CORP.										
DATE	CHK NO.	PAYEE	CHK AMT.	CONC.	PLUMB.	ELECT.	LUMBER	ROOF'G.	PAINT
7.9	917	A&D EXCAVATION	1,847.60	1847.60					
7.9	923	MARTIN CONCRETE	7,687.11	7687.11					
7.15	947	CITY STEEL CORP.	2,117.06	2,117.06					
7.16	952	DON'S PLUMBING	4,771.01		4,771.01				
7.17	971	STEVES PLUMBING	5,711.10		[illegible]				
7.21	984	A&J LUMBER	10,041.11				10,041.11		
[illegible]	[illegible]	MADISON'S INSUL.	1,653.87					1,653.87	
9.27	1004	MIKE & BOB'S ELECT.	6,304.11			6,304.11			
9.27	1021	JAMES ALLEN ROOFING	2,107.10					2,107.10	
10.3	1034	B&D PAINTING	478.09						478.09
—	—	TOTAL TO DATE	147,911.01	15,391.11	11,051.07	9,907.11	27,391.17	7,187.11	1,787.07

Job cost record
Figure 5-5

certainly payroll. Contractors hate paydays. They're always coming at the wrong times — like the beginning, middle, or end of the month, when there's no money left. As an employee, you couldn't wait for payday. As an employer, you want to put it off as long as possible.

Time Cards

First, let's take a look at time-keeping methods. Most companies use a simple time card that can be bought at a stationery store. See Figure 5-6.

The time card records the elapsed time each man worked. You can also use it to pin down the job and type of work done during that time. You could have a job number for each job, but some contractors use separate time cards for each job. I've found that it's simpler to use one time card for each employee and a separate job number for each job. The employee should write in, opposite the time worked, a two- or three-word description of the work done. This can be used later to figure costs and compile future estimates.

Generally, we use one card for each week. The time clock should stamp military time — hundred-minute hours and 24-hour days. That makes figuring each employee's work hours much easier. Your help costs money, not only in wages, but also in the time it takes you to look after their payroll and deductions.

Once you've figured the hours worked, I recommend that you have a payroll service figure the amount due each employee. A service will also figure tax and union deductions, of course; prepare returns for your signature; and issue checks on

TIME CARD

COMPANY NAME: GOOD-TIME CONST. CO. EMPLOYEE NAME: MIKE MACHO

DATE	JOB NUMBER	ACTIVITY	HOURS
MON.	C107	ROOF STRUCTURAL REPAIR	8.5
TUES.	C119	SECOND FLOOR FRAMING	11.5
WED.	C93	BUILT FENCE	7.0
THURS.	C93	FOUNDATION FORMWORK	9.5
FRI.	C93	" " "	8.5
SAT.			
SUN.			
TOTAL			44.5

SIGNATURE ______________________

Time card
Figure 5-6

your account. The cost of these services for a company with 5 to 10 employees getting paid twice a month will be about $50 per month. It should save several times that in manhours and frustration. Many independent services and banks offer a payroll service.

If you're using only subcontractors, there are no payroll deductions to worry about. But be careful. Having a contract with each tradesman or paying in cash doesn't make the recipient any less an employee. The IRS and your state figure that if you set the working hours, supervise the work, and use workers who don't have a business license or carry their own insurance, they're your employees. That lets the tax boys hit you years later for all the taxes you didn't deduct.

Withholding Tax

If you have employees, and you don't use a payroll service, you have to figure state and federal withholding tax for each employee each month. I'll explain how it works.

Not all states have the same withholding requirements. Your state will have a tax booklet that explains how to figure withholding for your employees. Note also that these rates vary both from state to state and from time to time.

Payroll withholding is detailed, exacting work. Failing to withhold or to forward amounts withheld can put you in hot water in a hurry. As an employer, you're always at fault when there's a mistake.

The tax money you withhold from employee checks stays in your bank account until the deposit date required by law. Deposit dates vary by state, and at the whim of elected and appointed officials. Here are some general guidelines. If total deductions are less than $500, your deposits must be made on the fifteenth day of the following month. If your total deductions are between $500 and

$2,000, your deposit will have to paid weekly. On anything above $2,000, you'll be required to make deposits within three days of issuance of your payroll checks. Until then, the money is yours to use. But don't come up short on deposit day! The actual deposit can be made at any bank that's a member of the Federal Reserve System. Your bank almost certainly is.

You make deductions for both the state and the federal government. FICA, or Social Security, is deducted until employee earnings reach the maximum set by law. The FICA deduction is about 7%. As an employer, you contribute an equal amount. The total FICA tax comes to a whopping $14 for each $100 of income. That's tough on the best of budgets.

FICA deposits made at your bank have to be accompanied by a deposit card issued by the IRS. The card is printed with your Federal Employer ID number, which you get from the local IRS field office.

FICA deductions are computed from the tables on Circular E, which you can also get from any IRS field office.

But that's not all: the state wants its cut, too. Most states have three basic deductions. I'm sure you've heard of the state income tax, state disability and unemployment compensation. State income tax is based on the amount earned and varies with the number of dependents declared. State disability is also based on income, but any income over a certain amount each year is usually not taxed. Unemployment compensation is based on gross company payroll. But companies with lower unemployment claims (a better *experience rate*) usually pay a lower rate.

If all this withholding seems too complex for you to waste time on, consider using a payroll service. We talked about payroll services earlier in this chapter. Many such services compute all withholding, make all deposits, and fill out the returns for your signature as part of their payroll service and at no additional cost. All you do, a day or two before paychecks are to be issued, is phone in the hours each employee worked. The service takes care of the rest.

Billing and Statements

Billing, or invoicing, must be done religiously. Billings and collections run on a thirty-day cycle in the construction business. Anything billed or paid sooner is done for accommodation only. The only exception to this unwritten law might be small, impoverished contractors who can't carry payroll on jobs that start and end in less than a month.

But don't let the thirty-day rule scare you. Nobody starts a construction company with that much financial cushion. In practice, you go from job to job trying to get far enough ahead to comply with the normal billing cycle. Some contractors finally make it and some never do.

All bills have to conform with the terms in your contract. We haven't talked about contracts yet, but we will later in the book. For now, let's just assume that you've got a contract and have done part of the work. Now it's the end of the month, time to get your bills out. Before typing the invoice, read the terms of the contract. It should spell out the date and amount you can invoice and the date you're to get paid.

Many construction contracts require that you submit an invoice on or before the 25th of each month for any payment due on the tenth of the following month.

Invoices for work in progress usually have a cut-off date. This will probably be the last Friday before the 25th. Pull all of the employee hours off the time cards through that last Friday. Then pull out your material bills, insurance, and any other costs you've incurred on that job in the last 30 days. Add the appropriate percentage for profit and overhead through the cut-off date. Your accountant can help you establish these percentages.

After totaling all this, tack on the charge for any authorized extra work you've done. Then, subtract out any credits or advances. Finally, total the amount owed at the bottom of the invoice. See Figure 5-7.

Statements are different from invoices. Use a monthly statement to summarize all unpaid invoices as of the statement date. Each statement should identify unpaid invoices by number and date. This is a reminder to the client: This bill is past due!

Make your statements as clear as possible. They should show the exact status of each account: the original contract amount, any net change in the contract due to extras and credits, the total amount that's been paid to the date, and the balance left owing after all payments have been deducted. See Figure 5-8.

INVOICE

COMPANY NAME: HILLTOP AUTO		DATE: 3.1.86
NO.	ITEM	AMOUNT
1.	HOURS WORKED THIS PERIOD 437 x $25.00	$10,925.00
2.	MATERIAL AND SUB CONTRACTORS:	
	A. PLUMBING	2,344.04
	B. ELECTRICAL	1,587.11
	C. FRAMING	4,701.53
	D. LUMBER	14,097.22
	E. CONCRETE	5,333.92
	F. MISC.	1,847.21
3.	LESS CREDITS FOR:	
	A. REDUCTION OF NORTH OVERHANG	728.00
	B. OMIT PANEL & PARKING LOT LIGHTS	3,411.00
	C. OMIT WASH DOWN SLAB	1,500.00
4.	EXTRA WORK APPROVED	
	A. ADD RECESSED LIGHTS	847.10
	B. ADD CABN'T. IN CONF. ROOM	525.89
	TOTAL AMOUNT DUE	$47,848.02

Invoice
Figure 5-7

STATEMENT

COMPANY: JONES AND BUSHMAN		DATE: 3.1.86
NO.	ITEM	AMOUNT
1.	AMOUNT OF CONTRACT	$187,402.19
2.	AMOUNT OF EXTRAS	32,112.88
3.	AMOUNT OF CREDITS	11,859.12
4.	CURRENT CONTRACT AMOUNT	207,655.95
5.	AMOUNT PREVIOUSLY BILLED	93,823.18
6.	CURRENT BALANCE REMAINING	$113,832.77
7.	AMOUNT PREVIOUSLY BILLED	$93,823.18
8.	AMOUNT PAID TO DATE	70,667.22
9.	OUTSTANDING INVOICES DUE:	$ 23,155.92
	A. INVOICE # 15076 ...... 11,791.03	
	B. INVOICE # 15092 ...... 6,707.70	
	C. INVOICE # 15181 ...... 4,657.19	
	D. TOTAL ABOVE INVOICES. . . 23,155.92	

Statement
Figure 5-8

You may have clients who delay payment because, they claim, your statements are confusing. Some will return the statement without payment but with a request for an explanation. Worse yet, they may just let the statement sit in a file for a month or two until you call. My advice is to make your invoices and statements as clear, complete, and concise as possible.

Some clients lose invoices or have a mailman who can't seem to deliver certain kinds of mail. Then you've got to deliver another copy of that invoice. This time, hand deliver it. Explain it to the client while you're at his home or office. Ask for a check before you leave. Don't leave any room for quibbling or excuses. No one goes out of his way to write a check.

Collections

Everybody hates collections, from the collector to the collectee. Collecting overdue accounts is rarely painless. Nobody wants to part with money if he can avoid it.

But getting paid is always up to you. Collecting is part of invoicing. Don't assume you're going to get paid just because you sent out a bill. Think of invoices as being like judgements in a lawsuit. It's up to the creditor to collect, even after the judge says that payment is due. Nobody's going to collect for you. You've got to do it.

The key to good collections is a combination of three factors. First, you can only collect from your client if he has the money in the first place. Therefore, collections have to begin even before you sign a contract. How? You investigate. Where's the money coming from? Is there a reasonable amount set aside for contingencies? Is your client financially solid? Can he pay his bills?

Second, collections are easier if you do your job conscientiously and according to the contract. If you're never on time, if your workmanship is marginal, if your budgeting is inaccurate, you can expect collections to be an Olympic event.

Finally, your billings must be in good order, on time, and have all necessary back-up materials enclosed. Otherwise, you'll confuse your client's accounting department, and get requests for further information and explanations rather than checks.

There's one more ingredient that can help when it comes to collections. It's known as persistence. Stay in touch with the people you invoice. Know where your payments stand. Know what's owed you and where it is in your client's accounting system.

Two weeks after bills go out, start calling accounts that haven't paid. Keep constant pressure on delinquent accounts. Apply the pressure and you'll be at the top of every payment list.

If your client needs a lien release or additional information, get it to him that afternoon. Don't let two or three days go by before you respond.

When an invoice has gone unpaid, or only partially paid, for more than 30 days, add interest at the maximum rate allowed under the contract and by your state. You don't expect to collect this interest, but it raises the settlement value of your claim as each month passes.

Some bills get paid sooner if you offer a 1% or 2% discount for payment within 10 days. But remember that a 1% discount for payment in 10 days rather than 30 days is about the same as paying 20% interest at an annual rate. That's an expensive way to do business if it isn't necessary.

If you have a tough time collecting from a client, repetitive calling may be the most effective collection technique. You'll either get paid or you'll find the reason why. At the very least you'll have his undivided attention.

Finally, if you expect that collection will be a serious problem, don't work a day past the payment date specified in the contract. Don't compound the problem by doing more work when you haven't been paid for work already done. If your client's going to stall a creditor let it be someone else, not you. Pull your crew off the job and get your lien papers in order. Let the client know just what to expect. He may never use you again, but at least you'll get paid for work already done.

Here's why a threat to walk off a job is usually very effective. Many owners have construction loans with draws against the loan every 30 days. But the draws can be delayed if there isn't normal progress at the job site. Threaten to hold a crew off the job and you'll get paid promptly nine times out of ten.

Some clients pay on time. Others would rather sell their grandmother than pay when payment is due, even if they have the money sitting in a bank somewhere. Every client is unique in some respect. You won't run into too many really bad apples. But one or two are enough to drive a contractor crazy.

Insurance

It's as easy to be insurance-poor as it is to be land-poor. Everything doesn't need to be insured. Accept some reasonable risks yourself. There's a limit to what any contractor can afford.

If you have disability insurance, use it only to supplement your Worker's Compensation Insurance. Coverage shouldn't exceed 75% of your monthly salary. The premium will be very small if there's a six or nine month waiting period before payment begins under the policy.

You have to carry Worker's Compensation Insurance to cover all employees, even clerical workers. Only employees are covered, not subcontractors. The cost is based on your payroll and the kind of work each employee handles. Unfortunately, the cost per $100 of payroll is very high for some construction trades. For roofers and steel erectors, Worker's Comp will cost about $20 for each $100 of payroll. The cost for carpenters is about half that. Clerical work is much safer. It carries a rate of about 50 cents per $100 of payroll.

Your insurance carrier has the right to audit your books quarterly to verify payroll and check the classification of all employees. Falsify a report and your insurance could be cancelled. But there are ways to save money. The classification system used by insurance carriers is fairly primitive. There's room for legitimate disagreement on what category a particular trade or job falls into. Your auditor will claim that each job falls into the highest-rated category. And he'll insist that division of each employee's time into several different classes is impossible or illegal. Good! You can play that game too. Arrange the responsibilities of each man on your crews to maximize the number of low-rated jobs and minimize the number of high-rated jobs. That's perfectly legitimate!

Here's an example. Truck drivers, supervisors and laborers carry lower comp rates than carpenters and roofers. If you have a four-man crew framing a house, one might be a full-time supervisor and another might be a laborer. Using two carpenters, a supervisor and a laborer will cost you $10 a day less for Worker's Comp than using four carpenters.

Distribute work so that some workers handle all high-rated tasks and others handle all low-rated tasks. Having all workers do both low- and high-rated work puts all workers in the highest-rated category.

Find out which jobs carry the lowest rates, and arrange your crews so that some workers handle only this kind of work. The insurance-rating bureau in each state publishes a book that describes the work each category of worker is assumed to handle. Your insurance agent can get a copy of this book and will quote the current rate for each category in the book. Assign work accordingly.

You'll need some insurance in addition to Worker's Comp. Every builder needs property coverage and public liability insurance. These are usually bundled in a policy titled *comprehensive general liability* and include coverage for vehicles, equipment, a fidelity bond, and umbrella liability coverage. The cost will usually be about $3 for every $100 of payroll. Any general insurance agent can supply details.

When shopping for insurance, call in two or three insurance agents who specialize in construction insurance. Tell them a little about your business, its volume and risks, and let them make coverage and cost proposals to you.

Once you've received their proposals, sit down and go through them to decide which coverage you can do without. Then buy insurance to cover the rest. The premium savings alone over a ten-year period can pay for a trip around the world for you and your wife. Here's a tip that could save a few hundred dollars a year. Your subcontractors should have their own coverage for many risks. It's cheaper to require them to buy their own insurance than to cover them with yours. Provide in your contract with them that they will carry auto and truck liability coverage.

Don't insure everyone and everything on the job. Insure just your risk and your people, equipment, and materials. Let others on the job cover their risks. In short, define your insurance risk and cover it. Then spread the remaining risk to those who should bear the responsibility.

Other Liability

Just having adequate insurance isn't enough. There are many ways to suffer a loss in the construction business. Insurance will cover many of these situations. But you're far better off to avoid losses whenever possible. And it's surprising how little effort is needed to prevent some major losses.

You can get hit with a loss any time from before the first nail is driven to ten years or more after the project's complete. Not every loss can be anticipated, much less prevented. But in today's litigation-oriented society, it pays to be on guard. Let's review a few of the situations that have high loss potential.

Your Own People: Employees promising more than they or you can deliver — usually without your knowledge — will get you in trouble every time. This is especially true if an employee has put a promise in writing. Prevent this situation by counseling employees on the risks of making any promise to anyone without management consent. Ask your attorney, insurance agent, or accountant to provide a few horror stories to make the point: Anything any employee does or says could be interpreted as a commitment of the company.

Working Conditions on the Job: Excess debris on the site gives the impression of poor supervision and invites suit if someone is injured. It also makes injury more likely. If materials don't fall on your tradesmen, your tradesmen will fall on the materials.

Poor control of access to the job site invites lawsuits. If a spectator or even a trespasser is injured, you'll waste hours or days in court and with attorneys. Post a sign telling potential trespassers to stay away. Make an occasional visit to the site between 5 P.M. and 10 P.M. (when most unauthorized access occurs), ask adjoining property owners to call you if they see anyone on the job after hours, and request tradesmen to inform you if they notice signs of unauthorized persons on the site.

Shoddy Workmanship: You know what poor workmanship looks like. Poor work that's uncorrected will eventually come back to haunt you. Don't ignore leaky roofs or windows, ductwork that won't stay together, concrete that cracks, and so on. Need I say more? Bad work is a bad deal for everybody. Only lawyers make money on it.

What about people in your organization who practice the fine art of shoddy workmanship? I'm not sure this can be cured. My advice: talk to them. If that doesn't do the trick, either give them a job they can't make a mess of or get rid of them as soon as possible.

Bonding

Bonding is something you do by choice. First, you decide what kind of builder you want to be. If you select certain kinds of work, you'll usually need to be bonded. To find out if you are bondable, call your insurance broker. He'll explain the requirements and the costs. If you can't afford the bond you need, you'll just have to go on cooking in a smaller kitchen until your finances improve.

Getting a performance bond is a lot like getting approval for a bank loan. The qualifications are about the same. If you plan on doing commercial or government work as a general contractor, you're going to need a bond on most jobs. If you're in poor financial shape, bonding may be impossible. You'll have to concentrate on residential jobs, or work for another contractor who is bondable.

Construction performance bonds guarantee that the contractor will finish the work at the agreed price. The need for bonding is obvious. Contracts go to lowest bidders, whether they can finish work at the price quoted or not. Too many contractors, stuck with what became a money-losing contract, just walked off the job. That leaves the owner of a partly-finished building looking for another contractor to finish the work.

Bonds solve that problem. The bond is written by an insurance company. It guarantees that work will be finished under the contract, even if the contractor abandons the job. Obviously, the bonding company has to have faith in the construction company it backs — faith that the contractor has the financial muscle and skills needed to finish the job.

The bonding company usually protects itself initially by requiring the contractor to keep at least 10% of the bonded amount in cash or securities on hand at all times. As time passes, they may reduce that ratio. But until they really get to know you and your operation, they'll keep a tight rein.

Here's an example of what I mean. Let's say you want to bond a $100,000 job. You'll need $10,000 cash in the bank. Later on, that amount may be reduced to $8,000, $5,000, or even $2,000, as you prove your ability to manage your work load and as you develop a track record.

Each time you bond, you'll have to update the financial statement on file with the bonding agent. The volume of bonded work can't grow unless your bonding capacity grows with it. If your capacity is $200,000 and you're doing a job for $180,000, you can only bid another job for $20,000. If you want to bid another $75,000 job, you'll either have to wait until your first job is nearly finished or get your capacity increased. If capacity can't be raised and you aren't close to finishing the first job, the only alternative may be to get the percentage reduced to less than 10%.

If you want to handle jobs that require bonds, build a strong relationship with an insurance agent who handles bonds. Your agent can intercede with the bonding company on your behalf. If he has faith in you and your business, bonds will be as simple as they can be (though they're not really very simple at all).

You may have to try several agents and bonding companies before you find one that appreciates your business. Not all insurance agents handle construction bonds. Deal with an agent who specializes in construction bonding.

Unless you plan to be the biggest contractor on the block, there's a limit to the bonding capacity you'll need. Look at it this way. Suppose you find that you're comfortable with jobs in the $100,000 range, and running two or three of them at a time is enough to keep profits flowing. Your bonding requirement is only $200,000 to $300,000. There's no way that you'll ever need a $500,000 or $1,000,000 bond.

As we discussed earlier, high volume usually reduces profit per dollar earned and increases complexity. So find your bonding niche and stick with it. Avoid the over-expansion mistakes that plunge contractors into financial ruin.

There aren't many ways to beat the bonding system. One thing you can't do if you want to get a bond is work on speculative projects. That's an absolute "no-no." Bonding companies are scared to death of involvement in projects like spec houses that are built out of the contractor's pocket. As far as they're concerned, if you want bonding, you'll have to forsake spec building completely.

If you want to participate in a speculative project and still be bondable, do it with a partner. You can do the actual construction, but the bonding company should understand that there's no risk on your part.

Suppliers and Purchasing

If your finances are in good shape, opening accounts is a snap. Most suppliers, however, will require a financial statement and several credit references from you before they'll make the first transaction.

If your finances are a shambles, getting credit is a different story. It's tough to work without credit, though it can be done. Most suppliers will want to put you on a C.O.D. basis. Fortunately, there are always a few who will extend anybody credit once. Suppliers who are new in business, or who need your account, will be more anxious to extend credit without a credit check.

Electrical and plumbing subcontractors have a tougher time getting credit without a careful credit check. Here's why. There are fewer electrical and plumbing supply houses than lumberyards. Each distributor knows nearly all the creditworthy plumbing and electrical contractors in the area.

If you can't establish credit, there are two choices. First, get a draw against the contract price before work starts and keep the payment schedule ahead of your expenses. Some owners and lenders will do this, either out of ignorance or as a favor to you. If that's impossible, ask your client to pay for materials as they're needed. Let the owner know before work starts that you can't finance material purchases out-of-pocket. He has to pay for materials when purchased.

Even if you have a good payment record, purchasing is never a routine and easy task. The easiest place to lose your profits is through lost or stolen materials. It's every contractor's responsibility to see that what's paid for actually gets used on each job.

There are hundreds of ways to lose materials. You've got to limit "shrinkage" to some irreducible minimum if you want to survive in construction. The most expensive stud on any job will be one bought to replace the one a carpenter took home with him.

How do you know that what was bought is really for your job? Did you do the take-off? Why are you short of studs or joists? Who checked to see that what was ordered was what got delivered?

I know these are hard questions. But let me ask you another. Do you want to make money, or are you just playing at contracting? If you can't keep track of your materials and your profits, you're running a charity, not a business.

I know a general contractor whose superintendent was ordering lumber, hardware, trusses, and other materials to build a mountain cabin. The only problem was that the project they were really working on was a five-hundred house subdivision in suburban San Jose, not a cabin in the woods. The contractor didn't discover that he was building

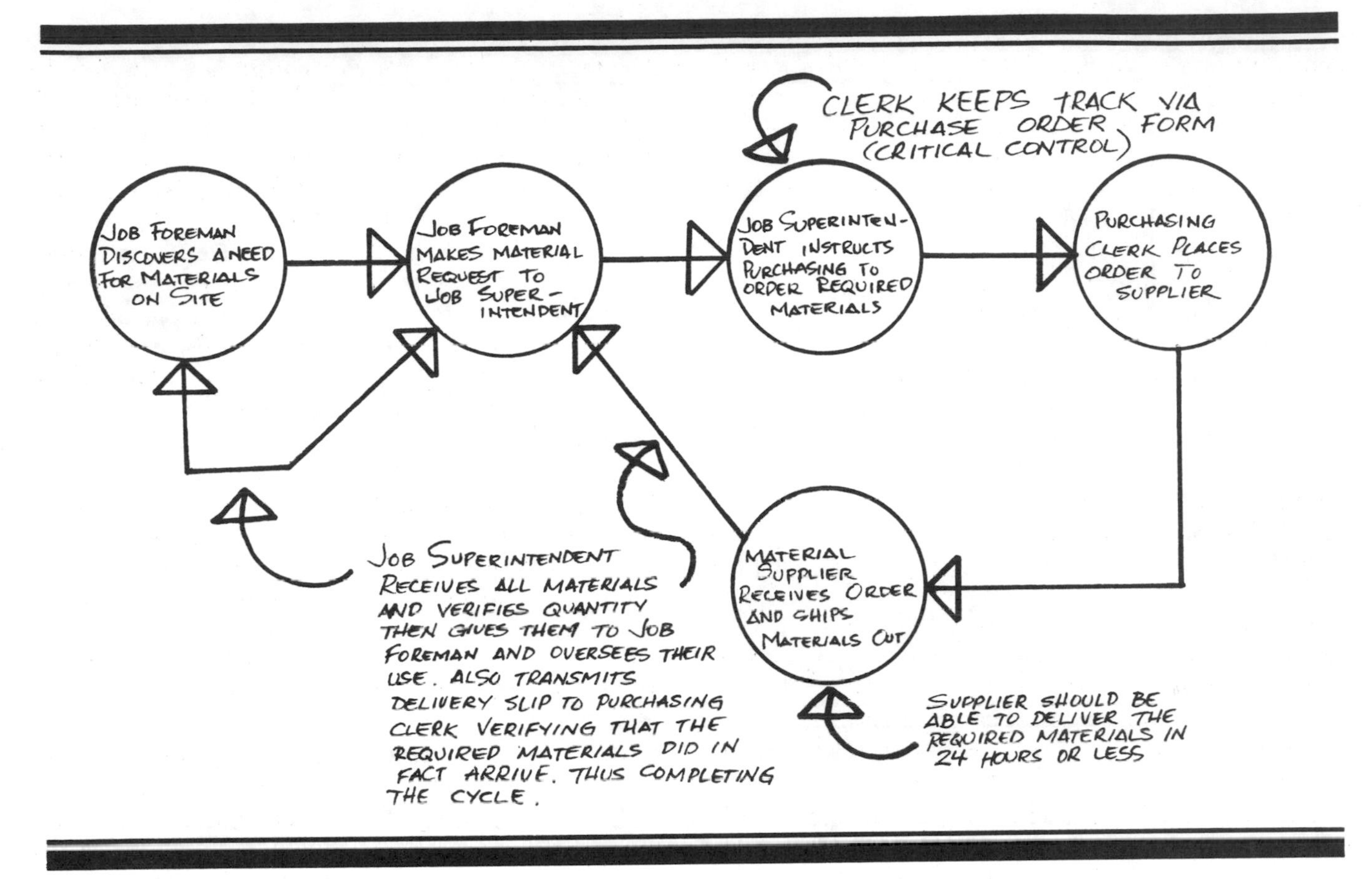

Purchasing process
Figure 5-9

a mountain cabin until he went out to the job site and reviewed the material lists personally. By that time, forty thousand dollars worth of materials had been diverted off the job site.

Control of purchasing should be as routine and simple as possible. Figure 5-9 details a purchasing process. If your company is small enough, handle purchasing yourself. If not, confine it to two people: the superintendent and the purchasing clerk.

Here are the steps involved in getting any piece of material to the job site: First, a tradesman, crew leader, or foreman discovers the need for an order. The foreman makes a request to the superintendent. The superintendent instructs the purchasing clerk (or secretary) to order the materials required. The purchasing clerk places the order. This must be done with a properly completed purchase order, or you'll get lost in your accounting, and your billing won't come out right.

If the supplier's dispatcher is on the ball, you'll get a 24-hour delivery. If you need same-day service, have a courier pick up the item. The superintendent receives and signs for the materials requested. He compares what was ordered with what was actually received, and informs the purchasing clerk of the materials that arrive each day.

This procedure may seem cumbersome, but it's not nearly as expensive as re-ordering lost or misplaced materials. With a little practice, nearly every step in the purchasing cycle — except actual delivery — can be done by phone.

Purchase orders must be simple. You'll need a form for your superintendent to fill out when he orders. It should show the job number, the date, what's being ordered, the quantity, a delivery date, and the job address. There should be a space for a signature confirming receipt of the order. See Figure 5-10.

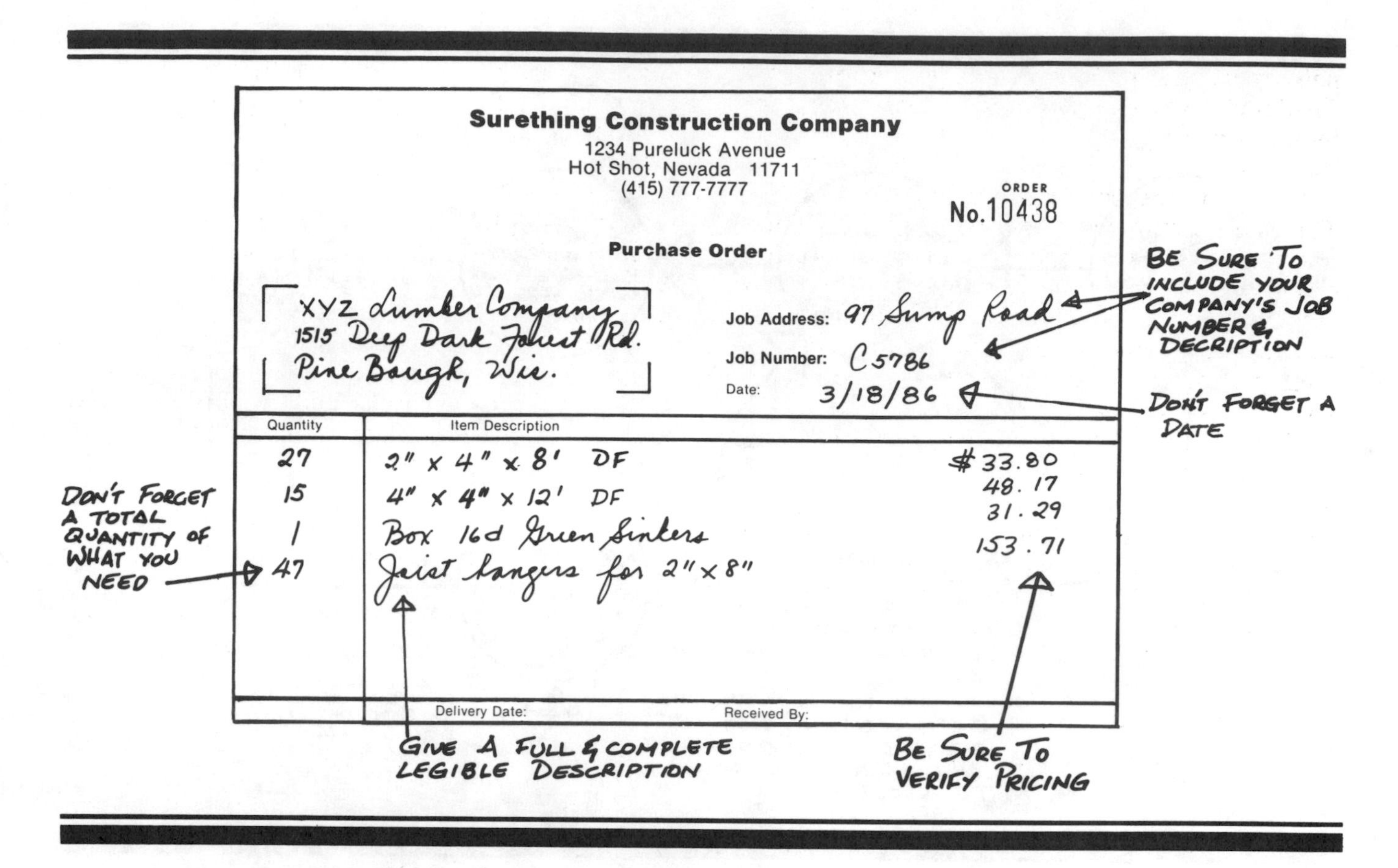
Surething Construction Company
1234 Pureluck Avenue
Hot Shot, Nevada 11711
(415) 777-7777

ORDER
No. 10438

Purchase Order

XYZ Lumber Company
1515 Deep Dark Forest Rd.
Pine Bough, Wis.

Job Address: 97 Sump Road
Job Number: C5786
Date: 3/18/86

Quantity	Item Description	
27	2" x 4" x 8' DF	$33.80
15	4" x 4" x 12' DF	48.17
1	Box 16d Green Sinkers	31.29
47	Joist hangers for 2" x 8"	153.71

Delivery Date: Received By:

Purchase order
Figure 5-10

Your purchasing clerk uses either a similar form, or a standard accounting form that you can get at many stationery stores. The difference in the clerk's form is that there should be room for a purchase order number and for some way to note whether the materials are under the basic contract, or are extras that must be billed separately.

No matter what purchasing system you follow, here's a suggestion. Don't take delivery of any material that can't be locked up or installed by sunset. Materials have a way of walking off the job between quitting time and 7 A.M. the next morning.

Equipment

I think half the equipment I buy has legs: it's always walking off the job. And the best stuff seems to have the best legs. Broken equipment never leaves.

Every contractor will admit that good equipment is hard to keep track of. The best control method I've found is the old "ball monitor" routine we all learned in elementary school. Start by assigning an identification number to each piece of equipment you own. Keep the ID numbers in a master record, and keep the equipment locked up whenever possible. When it's not possible, be sure one person has responsibility for checking the equipment out and checking it back in. Checkout authority might rest with a yard foreman, a purchasing clerk, or a secretary. Anyone who needs equipment or supplies checks them out with the responsible person. The individual in charge physically checks the equipment out to someone by name, date, and serial number, and notes the condition of the equipment.

Equipment rental yards operate this way. Do this and you at least have some idea of where your equipment is. If something's missing, you know who to see.

Keep in mind that the cost of missing tools, nails, gas, or materials reduces your profit dollar for dollar. Any loss avoided increases your profit by that much.

Embezzlements and Forgeries

This is an unpleasant topic. But I'm not going to duck it. Let's get it right out in the open: Some employees will pick you clean if they can get away with it. I'm not just talking about occasional workers or subcontractors you hardly know. I'm talking about trusted regulars who have been on your payroll for years.

This is the voice of experience speaking. I know about embezzlements and forgeries. I was a victim to the tune of $15,000. An employee in my office gave me a real education in the finer points of checkbook gymnastics. I didn't know you could do some of the things this person did with my checkbook. It was an expensive lesson, and one you should avoid.

Keep your own checkbook. No matter how profitable you are, you're giving away the combination to the safe if anyone else reconciles the account and reviews the cancelled checks.

Even if your bookkeeper doesn't sign checks, embezzlement is possible. I'm the only one who signs on my company checking account. And still I was taken to the cleaners.

The best protection against forgery is to keep your checks under lock and key and to review all cancelled checks returned by the bank. Beware if checks are missing from the back of the checkbook. The forger may be hoping that the theft will go unnoticed for several months.

If you are a forgery victim, present your claim to the bank. They're responsible for clearing forged checks. You can sue them for restitution.

Consider buying a fidelity bond on your bookkeeper. The cost is modest and any insurance agent can provide coverage. You could pass the cost of the bond on to the bookkeeper as a condition of employment.

You could also buy a check-writing machine, from a company that offers insurance protection against most of the alterations a potential check forger might make. Again, the cost is modest compared to the possible loss.

The easiest way to spot a forgery is to review your signature on all cancelled checks. Most people find it easy to recognize a forgery of their own signature. A signature forged by an amateur will probably have a faint pencil outline under the actual inked signature. That's a tip-off that there's been foul play.

Avoid signing checks with a broad felt tip pen. That obscures many of the small characteristics that make your signature distinctive. A ball point pen emphasizes the pressure points in your signature. That's an important characteristic — and it's difficult to copy.

Here's another hint. Sign your name clearly. The sloppier your signature, the easier it is to copy.

Sign your own checks and reconcile your account every month. Match your check stubs, or file copies, with the checks themselves to see that they're the same. Then tally the checks against the monthly bank statement. Count the cancelled checks and count the check debits on the statement. Finally, have your accountant (or someone other than your secretary or bookkeeper) go over the ledgers at least once each quarter.

Embezzlement is easy for someone who knows the system and who works without direct and constant supervision. Let me give you some examples.

Have you ever asked your secretary to pick up $50 at the bank for the petty cash fund? You ask her to write a $50 check. She brings you a dated check made out for $50. You sign it and ask her to go to the bank, cash it, and bring back the $50. Does that sound familiar?

Here's what might happen. Unknown to you, she puts a *One* in front of the *Fifty*, writes $50 on the check stub, then goes to the bank to cash the check. She puts $100 in her purse and the remaining $50 into the petty cash fund.

When the statement comes in, she destroys the altered check. You're happy. The check stub looks right. The incriminating check is gone forever. She enters a $50 notation in the check ledger at month end. Unless you (or your accountant) reconcile the statement carefully, that $100 becomes just an unexplained error that's adjusted at year-end. No one will ever find the $100! Pretty slick? Could it happen in your business? Of course it could!

Here's another gambit. A regular supplier calls to report that some materials are ready for pickup. Your secretary takes the call and notes the amount due as $654.29. But she makes out the check for $754.29 and you sign it! You send the secretary to pick up the goods. She comes back with three things: the order, an illegible or altered receipt, and

$100 cash in her purse. You've used this supplier many times and trust him. The cost is always reasonable, so you don't check the invoice. The result: your secretary gets an extra $100 in tax-free cash, because she's so friendly with your supplier. Besides, your checks are always good.

Here's one that's even harder to protect yourself against. Say your company name is First Rate Construction. What's to prevent someone else from opening a bank account in that name? Nothing. Anyone who files a "DBA" (Doing Business As) Statement with the County Recorder can open a bank account — in the name of Satan if he wants to.

Now what's to prevent someone in your office from slipping a check or two out of the incoming mail, endorsing it with the phony DBA, and depositing it in a new account? Nothing at all! Once an account's opened, it's an easy matter to deposit checks, then to withdraw cash a few days after the check clears. If that employee also has access to customer payment ledgers, this little theft could go on for years before it's discovered.

The Labor Relations Board

Most states have a Labor Relations Board that's authorized to hear employee complaints against their employer. The law usually provides that employees have to be paid within a certain number of days of the end of a pay period. If they aren't paid, they can complain to the Board.

A financially-troubled contractor can learn to hate the Labor Relations Board. They act like a free attorney for your employees and double as the judge during the hearing.

The Labor Relations Board will teach you to fire employees the minute it looks like you can't meet a payroll. The worst possible thing you can do is to hold onto your help, hoping that you'll be able to pay them later or at the end of the month. Once the grace period has passed, you can guess what your employees' next stop will be: the Labor Relations Board.

All employees have the right seek help from LRB if they aren't paid. After all, you're the one that failed to perform. It's no defense that their work was shoddy or that they made costly mistakes. The only issue is whether they worked and didn't get paid.

The LRB will probably assess fines against you for unpaid wages. These fines are based on the salary of the employee involved. The bigger the salary, the larger your fine. A typical LRB penalty would be $50 per employee for each day they remain unpaid. That can add up fast for a crew of five framers.

The penalty assessment is supposed to teach employers to make payrolls on time. What it really teaches contractors is to dump help the minute making payroll looks unlikely. Hiring new people later is a better bet than battling existing employees at the LRB. Dump and rehire is the name of the game today.

The LRB teaches contractors not to bet on improving economic conditions. If you bet wrong, the penalties will be heavy. Because of this, many contractors dump and rehire help regularly. That's not the best way to run a company, but it's what the law encourages.

If you can't make next Friday's payroll, lay everyone off now. If you can't pay them in full immediately, pay them something each week until the date of your hearing. Here's why. LRB penalties are computed from the date of last payment. The board doesn't inquire into the amount of the last payment. Even relatively small payments will cut the penalty substantially. That could be important during a recession, when more employee claims are filed with both the LRB and the courts. It may take several months for your case to be heard.

Here are a few pointers if you're facing an LRB hearing. First, prepare your case carefully. Bring all relevant documents to the hearing: time cards, contracts, and any other agreements with ex-employees. Review your records carefully so you understand exactly what they mean. Don't rely on your secretary or bookkeeper to make your case. It's not their problem — it's yours.

Remember, any loss due to an employee error is irrelevant to the LRB. It has nothing to do with any wages claimed. If you feel that an employee has cost you money, take your case to the appropriate courthouse. The LRB doesn't handle that kind of dispute. Their sole function is to determine what wages are owed.

If the Board decides against you, here's what to expect. A judgement is entered into the public record against the employer, either you personally or your company. The LRB will provide the employee (claimant) with an attorney if the employee wants to pursue you. Of course, the cost of pursuit will be added to your fine.

If your ex-employee doesn't pursue the collection matter, the judgement becomes a lien on any house or other property your company owns. When you try to sell, the lien shows up and you have to pay.

As you can see, losing at the LRB is just like losing in court. Payment isn't automatic. If you're in desperate financial condition, you can always bankrupt your way out of the obligation.

Summary

Keep good records. Accounting is essential to success in the construction business. Balance your checkbook every month. Keep a cash journal, a client record, and a check register.

Payroll is exacting work. Be careful with deductions and withholding. Mistakes are costly. You'll pay both the tax and penalties for many errors.

Do your billing on a regular basis. Remember, collecting is part of construction contracting. You have to go after profits. If an account is delinquent, don't let more than two weeks go by before calling for payment.

If you're bidding a project that requires a bond, find yourself an insurance agent who knows bonding. Remember, bonding companies frown on spec building. Keep spec building a separate operation.

Only a fool assumes that every employee is honest. If your accounting staff is inventive and cautious enough, they can nick you four or five times a month in all sorts of ways. This is embezzlement, no matter how little is taken. Theft is theft. Do your own purchasing. Control the use of materials yourself. Write your own checks. Sign your name clearly. Finally, have your accountant — not your secretary or bookkeeper — go over your ledgers as a matter of routine.

Who, Me Work?

"What I want to do is buy a business, then sit back and let the employees do the work while I rake in the money."

Have you ever heard anyone make a remark like that? It's crazy. Someone who says that has never run a company. When you have employees, you can't relax. Employees can make expensive mistakes, eat up the profits by milking their jobs, and do everything but what's needed to get the job done. No matter how good your business is, you can't sit back. Every entrepreneur has to be his own hardest-working employee if the business is to survive.

Sole Proprietor or Partnership

At one time or another, you're going to consider taking on one or more partners. Operating as a sole proprietor can leave you too short of talent and with too little cash to tackle the larger, more profitable jobs.

Partnerships sometimes work well when partners complement each other. For example, one might have design, engineering, or accounting skills. The other might be an experienced supervisor, estimator, or salesman. There are probably as many reasons to form a partnership as there are potential partners in the construction industry. But there are also some disadvantages to partnerships.

The danger in a partnership is that each partner loses some of the incentive to make the company succeed. Each participates only partially in the profits and shares only partially in the losses. It's like dogs pulling a sled. Sometimes you can't tell who's doing most of the work and who's just going through the motions.

The more principals a business has, the greater the chance that some partners are not pulling their share of the load. With many partners sucking cash flow out of the business, work has to be done very efficiently indeed. Many partnerships end up going broke or breaking up just to avoid bankruptcy. See Figure 6-1.

There's nothing quite so efficient as a unit of one. If you're a sole proprietor contractor, chances are good that you'll cut waste and overhead to the bone. You may be 50% to 80% efficient. Add one partner and you both may be operating at between 40% and 60% efficiency. With two additional partners, odds are that you'll only hit 20% to 40%. Add a few unproductive employees and the biggest worry may be meeting the next payroll.

Here's the key to successful partnerships, in my opinion. It guarantees that each partner pulls his share or suffers the consequences. Make each partner's income proportionate to the income he pro-

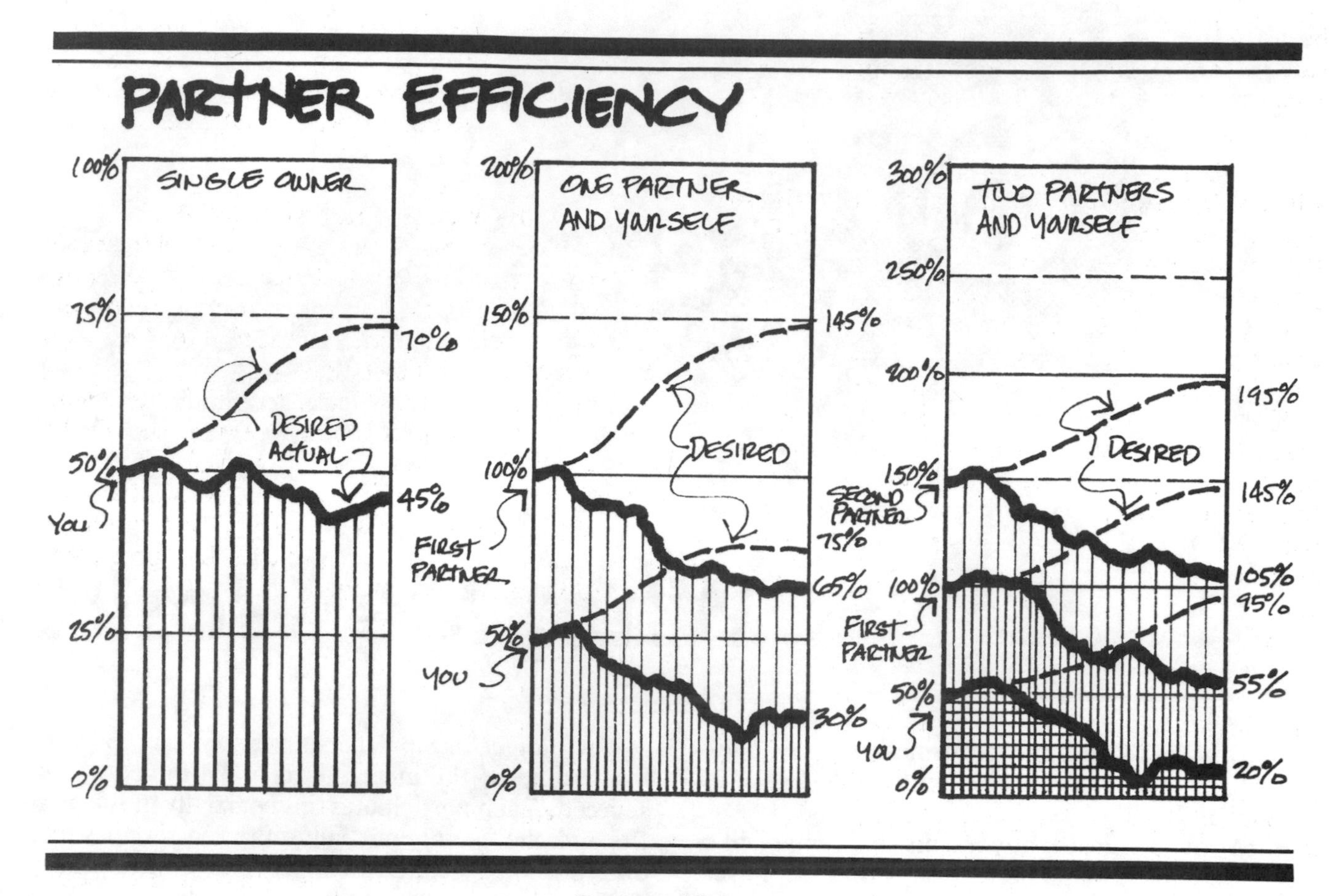

Partner efficiency
Figure 6-1

duces. How do you do that? Easy. Lay down a rule at the beginning that at least 50% of each partner's time should be directly billable to clients. Every partner should record time spent on every job, and that time should show up in the charges against some client.

Of course, the client doesn't have to know that he's paying for a certain number of hours of a particular partner's time. But it should be recorded in the partner's books. If some job isn't producing revenue for the partnership, the income for the partner or partners working on that job should reflect the loss. That way, there's likely to be very little wasted effort and time. If there are more than two partners, the ratio of billable time should be higher than 50%.

Make every partner's income proportionate to the income he produces for the partnership. That way you'll have no trouble with unproductive partners. Let partners work on non-billable projects and hire unproductive employees, and you're going to attend a lot of prayer meetings: partners gathered around the conference table wondering where the money for the next payroll will come from.

If you're going to do business with one or more partners, get started out right. Begin with a set of common goals. If you and your partners aren't heading for the same destination, you'll waste too much time pulling in different directions. Put your goals and expectations on paper. Refer to them occasionally. Revise them when necessary. Keep an eye on these goals and judge your success or failure by them.

And remember that a partnership is a lot like a marriage. Get to know the spouse and family of prospective partners before signing a partnership agreement. Your partner's spouse could turn out to be an ally or an adversary. Remember, anybody can get along when business is good. It's when business is a struggle that partnerships split up.

Job Supervision and Your Foreman

Big construction companies can afford foremen who don't work with tools. On a $10,000,000 job the foreman doesn't have time to drive nails. But you won't see many jobs like that. If you were a multi-million dollar contractor, you wouldn't be reading this book. For most builders, a non-working foreman is a luxury to avoid.

Your foreman should never be a spectator. Contractors with straw bosses don't make money in residential construction. Every construction entrepreneur has to be efficient. Efficiency begins with you and extends through your foremen to every man and woman on the payroll. A foreman can work as a framer, plumber, or electrician without sacrificing his efficiency as a superintendent. Leading by example is the most effective form of leadership. Don't tolerate a foreman who thinks working with tools is beneath his dignity.

The best kind of foreman is the expediter — the man who's determined to do professional work in the shortest possible time. He ensures that all tools and materials are in place and ready to go at the start of the day. He's ready to overcome any obstacle that might slow up the job. He can tackle any part of the work that needs his attention. He's always one step ahead of his crews, planning what they'll do next so the tools and materials are ready when the men are ready for them. He's willing and able to chase down materials, do take-offs, keep time cards, coordinate trades, pour concrete, tie steel, dig trenches, and in short, keep the job running smoothly and profitably.

Your foreman should understand that he's there to make money for his boss, not the other way around. If your foreman's there only until he can go off to run his own business, you probably have the wrong man for the job. And don't tolerate a foreman who's a plodder.

Finding the right man, and keeping him, is a tough problem. It could take a couple of years and six or eight trials before you find the right one.

Don't hire any foreman until you really need one. A three- or four-man crew doesn't need a foreman. The minute you hire one, your profitability drops. Don't take on help unless, and until, you're sure you need it.

Here's a final point on supervision. Don't assume that hiring a foreman relieves you of responsibility for supervision. Somebody still has to supervise the supervisor. Nobody has your interests at heart quite like you do.

Expecting the Unexpected

I could name at least a thousand things that can go wrong in the construction business — every day. And part of construction contracting is preparing yourself emotionally and financially to deal with these problems as they come up.

Experienced builders learn to handle problems routinely. They expect the unexpected, and find a way to get the job done and to get paid no matter what obstacles come along. But no construction contractor will survive long enough to become an experienced "pro" if he can't beat the most common construction problems:

1) Contracts
2) People
3) Materials
4) Insurance, Bonding and Taxes
5) Accidents and Mistakes
6) Payments

The biggest contract problem will always be nonperformance: Your client can't, or won't, do something he promised to do, or claims that you haven't done something you promised to do.

People problems are the most exasperating. Your foreman quits in the middle of the busiest month, your framer breaks his hand, your bookkeeper quits to get married, your plumbing sub threatens to burn the job to the ground.

Materials get lost in transit. If you don't get shorted, you're sure to get the wrong items, broken goods, or second-class merchandise. If some manufacturing plant doesn't go on strike, one of your suppliers will probably go out of business altogether.

Beating material problems is relatively easy if you use several suppliers for most common materials. Pay only the minimum amount with each order. If your primary supplier closes up, there's still money available to reorder elsewhere.

Your insurance and bonding agent can supply some nasty surprises: Rates can go up unexpectedly. Your bonding limit could be reduced without warning. A major loss could be uninsured because of a loophole in a policy. After all, who bothers to read insurance policies?

My advice is to overestimate your insurance and bonding costs. Get a good insurance man who specializes in contractor coverage and bonding. Then give him enough business so he recognizes you as a good client.

The tax man will leave you alone if you take taxes seriously and pay them on time. Getting behind in withholding taxes is real trouble. Your business might be locked up; your cars, trucks, and equipment impounded; and your receivables seized. It's not smart to fool with Uncle Sam.

An accident or a mistake could kill you or one of your workers. A well-run job is a safe job. Keeping everyone informed will prevent many mistakes. Consider having an occasional job conference or progress meeting. It may keep the plumber from digging up the electrician's power line, and the cement trucks from driving over a newly-laid drain line. Keep your job site picked up and your materials organized and you'll prevent a lot of accidents and dumb mistakes.

Finally, start planning to get paid from the minute you take on any job. Don't assume that you'll get paid just because you did the work. You have to work at getting paid. Stay in touch with your client and his accounting people. Get your bills in on time, or early if possible. Keep a record of your client's checking account number and know where he does his banking. When your check's ready, go pick it up. Don't use the mail. That's another chance for something to go wrong.

Losing Time

Let me give you a short lesson in waste. I used to work for one of the country's largest architectural-engineering companies. They had 1200 employees. I estimate that each employee wasted an average of 15 minutes a day. The cost is figured as follows: 15 minutes times 1,200 employees equals 300 hours a day wasted. Multiply that by 5 days a week, and you get 1,500 hours a week. Now multiply by 52 weeks and you have 78,000 wasted hours per year. Multiply by the billing rate of $35 per hour to get $2,730,000 a year in unbillable time. Now, if the average wage is $10 an hour, the real outflow of cash to payroll is an additional $780,000. That's $3,510,000 annually in real money lost. Keep in mind that we're talking about wasting only 15 minutes per employee each day. It's easy to waste that much time.

Figure 6-2 is a Time-Waster Chart. It should give you some idea of where you stand on this issue, and how devastating wasted time can be to your business. Let's bring an example of wasted time down to numbers every contractor can appreciate. Say you have five employees and each one spends an extra 30 minutes a day on coffee breaks. The annual cost to you will be roughly $15,000 a year if employees work at a $25 per hour charge-out rate.

Maybe your crews don't take an extra 30 minutes for coffee. But that's not the only kind of wasted time. Does the lumber company drop its load on the far side of the lot so your crew can spend the next two weeks manually hauling studs and joists 50 yards? It happens all the time. How about placement of sand and gravel? Does your roofing supplier drop his shingles at the curb line instead of on the roof? Did the heating and plumbing subs hack up your framer's work? Does the duct work have furred space to run through? Is everybody sitting around waiting for you to solve problems like these — problems you should have anticipated?

Think about it. The meter's running all the time. Somebody has to pay.

Another spectacular time-loser is poor ordering and haphazard delivery scheduling. Lack of materials leaves your help standing around, unable to do their work. Every $25 lost to wasted time reduces your profit by $25. And it takes about $500 in construction work to make $25 in profit. So wasting $25 in time is like losing $500 worth of construction work.

Waste of equipment is nearly as common as waste of time. I've seen contractors buy backhoes and tractors and then let them sit unused for months. That's plain stupidity. Did you ever rent a forklift and return it one day late? That could add a full week's rental charge. Where's your profit now?

Don't let things like that happen in your company. Construction takes careful planning and scheduling. Spend the time it takes to run your business efficiently. That's the best way to save both time and money.

Risk Management

One job in ten is a money-loser, no matter who handles it. Try to spot that job before it gets to you, and then avoid it like the plague. Here's a rule I follow: *You can't make money from people who are broke*. It isn't always easy to be sure, but you'll

TIME WASTER CHART

	AMOUNT OF TIME WASTED ON NON PRODUCTIVE ACTIVITIES		NUMBER OF EMPLOYEES (LOSS IN HOURS)								
			1	2	5	10	20	40	60	80	100
1.	COFFEE BREAK	.25 HR.	.25	.50	1.25	2.5	5	10	15	20	25
2.	REST ROOM	.25 HR.	.50	1	2.5	5	10	20	30	40	50
3.	SLOW XEROXING	.25 HR.	.75	1.5	3.75	7.5	15	30	45	60	75
4.	SLOW WORKER	.50 HR.	1.25	2.5	6.25	12.5	25	50	75	100	125
5.	LONG LUNCH	.50 HR.	1.75	3.5	8.75	17.5	35	70	105	140	175
6.	EXCESS TRAVEL	.50 HR.	2.25	4.5	11.25	22.5	45	90	135	180	225
7.	GEN. GOSSIPING	.25 HR.	2.5	5	12.5	25	50	100	150	200	250
8.	MISTAKES	.50 HR.	3.0	6	15	30	60	120	180	240	300

NOTE: THE TIMES IN ITEMS 1-8 ARE ADDITIVE, SO AS YOU ADD MORE WASTED ACTIVITIES, THE TIME EXPANDS RAPIDLY. I.E. IF YOU HAVE 20 EMPLOYEES EACH WASTING 1.75 HRS EACH DAY x 250 WORK DAYS = 8750 LOST HRS. x $35.00 = $306,250 LOST INCOME ANNUALLY.

Time waster chart
Figure 6-2

develop a sixth sense about this eventually. If the guy you're having lunch with seems as broke as you are, let him buy lunch — and then find an excuse to leave promptly. If you're lucky, he won't build anything. If he does, he'll do it a piece at a time. Your chances of getting paid are somewhere between slim and none. He's wasting your valuable time.

Risk can take many forms. Some jobs are unsafe for the tradesmen who have to work on them. Others have financial risk. Some simply require more skill, more equipment, or more financial muscle than you possess.

Know your limits. Contractors who exceed their limits end up in trouble. Many contractors have taken heavy losses by attempting projects that were either too big or too complex.

Many real estate developers have learned to spread their own risk to the contractors they use. Some of these developers start a project with full knowledge that there isn't enough money to finish it. They expect to sell out before construction is complete. If the project doesn't sell, the last contractors or subcontractors to finish work don't get paid. That's why it's more risky to be a drywall, paint, tile, or carpet contractor.

There's nothing wrong with risk, of course. Many builders and contractors thrive on it. But it's important that you evaluate the risk before accepting the job. Along with higher risk should go higher potential reward. If the rewards aren't there, don't accept the risk.

The ideal client brings very little risk. Look for well-financed corporate clients who must build because their business is growing. The electronics, energy, recreation, and food industries have had good prospects for growth over the last 10 years. Find clients in these industries and grow as they grow. That's a formula to help you thrive in the building business.

Summary

A sole proprietorship is probably more efficient than a partnership. Every time you add a partner, efficiency drops. A partnership works when the partners have similar goals and some objective measure of each partner's performance. Distribution of income should be proportionate to performance.

Your foreman, if you hire one, should both work *and* supervise. He should be willing to chase down materials, do take-offs, pour concrete, do framing — whatever has to be done to finish the job.

Be prepared for accidents, misunderstandings, hikes in insurance rates, strikes, late deliveries, and difficult collections. It's all part of the construction game. Preparation, constant checking, and careful planning will minimize the impact of surprises.

Lost time chisels away at your profits. Watch those coffee breaks. Use the Time-Waster Chart, Figure 6-2.

Risks? They're everywhere. Some jobs aren't your cup of tea. Ultimately, the best protection against risk is your own good judgement.

The Equipment Payment's Past Due

As far as I can tell, most contractors are working so they can be financially comfortable. Everyone wants enough of the things in life that cost money. But money isn't everything, even in the construction business. Having the time to enjoy life is just as important. And that's the root of a major conflict.

For most contractors, there's a tradeoff between time and money. To become financially independent you have to work like a dog, giving up much of the free time that others enjoy. But, spend too much time in unproductive leisure and you may never reach financial independence.

Keeping our noses to the grindstone right up to retirement day isn't the answer either. By then, we've lost our teeth, our sight, and our hair. We're probably too old and too tired to enjoy the money and time off we've earned.

Here's my philosophy. It may help you understand why I keep my business as simple as possible, and why I'm not interested in becoming the biggest contractor in the state. My view is that obligations and commitments can make you a prisoner of your own business.

Here's an example. Everyone wants to drive a nice car. The owner of a successful and growing construction company should drive a late model, flashy car, right? Of course! So you lease a car for $500 a month. That's $6000 a year that comes right out of profits. If profit averages 10%, you have to do 10 times that, or $60,000 extra every year, just to make the lease payment. How many extra hours of time are needed to generate $60,000 in business each year? My guess is that it's more than a few Saturdays, nights, and holidays.

But that's just one obligation. Consider heavy equipment, swimming pools, second cars, office space, trucks, saws and tools, phone bills, taxes, insurance premiums, accountants, lawyers, foremen, secretaries, workmen, and office furniture. Can you see why there's never enough free time?

If you ever expect to have any time off, short of dropping dead, you're going to have to decide which obligations to accept and which to duck. Begin by making two key decisions: What kind of contractor do you want to be? How big do you want to get? Figure 7-1 shows some of the tradeoffs between time and obligations.

There's no getting around it. Every construction contractor has obligations. But some kinds of construction involve more and heavier obligations than others. Let's look at a few examples.

EXPENSES VS TIME OFF

NO.	EXPENSE ITEM	MONTHLY PAYMENT	ANNUAL PAYMENT	ANNUAL VOL. TO MAKE PAYMENT	HRLY RATE	HR. VOL. TO MAKE PAYMENT
1.	HOUSE PMT.	$1000.00	$12,000.00	$120,000	$40.00	3,000
2.	FOOD PMT.	600.00	7,200.00	72,000.00	40.00	1,800
3.	AUTO PMT.	250.00	3,000.00	30,000.00	40.00	750
4.	UTILITY PMT.	400.00	4,800.00	48,000.00	40.00	1,200
5.	POOL PMT.	300.00	3,600.00	36,000.00	40.00	900
6.	BOAT PMT.	150.00	1,800.00	18,000.00	40.00	450
7.	R.V. PMT.	250.00	3000.00	30,000.00	40.00	750
8.	SECOND CAR PMT.	200.00	2,400.00	24,000.00	40.00	600
9.	CABIN PMT.	650.00	7,800.00	78,000.00	40.00	1,950
10.	INSURANCE PMT.	150.00	1,800.00	18,000.00	40.00	450
11.	HAWAII VACATION PMT.	400.00	4,800.00	48,000.00	40.00	1,200
12.	TOTALS	$4350.00	$52,200.00	$522,000.00	40.00	13,050

ITEM #1, HOUSE PMT = $1,000.00 X 12 MOS. = $12,000.00 = 10% PROFIT REQ'D. TO MAKE PMT. THEREFORE $12,000.00 X 10 = $120,000.00 ACTUAL CONST. WORK REQ'D. TO PRODUCE THE 10% $12,000.00 HOUSE PMT. ALL PMTS. REQ. 13,050 HRS. OF CONST. WORK.

Expenses vs time off
Figure 7-1

Capital-Intensive Contractors

In our business there are two extremes: *capital-intensive contractors* and *paper contractors*. If you own a construction yard full of equipment, supplies, and inventory, you're a capital-intensive contractor. Chances are excellent that you're becoming a slave to that investment.

In a capital-intensive operation, the owner (or owners) is preoccupied with making the next payroll, putting out fires, and meeting with bankers, accountants and lawyers. If he's lucky, he gets two weeks off a year. Most aren't that lucky.

Don't get the wrong impression. I'm not against owning equipment, or against running a capital-intensive construction operation. I'm against the strain that this sort of operation puts on you and your family. It takes a lot of "gravel" to be the head of a big construction company.

There's a problem that lurks just below the surface for every empire builder in the construction industry. The incredible cost of the equipment, the labor, and the financing are a constant drain. When operating a big company, you're forced to take jobs you don't like. An insatiable need for cash flow forces you to work with people you don't trust. Business pressure probably requires that you abandon all self respect just to keep the company afloat.

General contractors with loaders, hoes, and trucks are working slaves for their lenders. Supporting heavy payroll makes you a slave to clients who deserve neither your respect nor your time. Disputes with irate owners, lenders, employees, and other contractors are inevitable. The result? You spend countless hours settling problems rather than building.

If you're always chasing payroll, and fending off hostile clients and loan officers, *stop*. There are other ways to make a living in the construction business. Don't get caught up in an endless merry-go-round of loan payments and payroll.

Paper Contractors

Here's my favorite description of a paper contractor: *A builder whose major piece of equipment is his license.* Paper contractors don't exactly work without people and equipment. It's just that they severely limit that kind of involvement, and they hand-pick their jobs.

The nice thing about paper contracting is the low overhead. The insurance cost is minimal, payroll is small, loan payments and taxes are insignificant.

As a paper contractor, you can pick and choose the jobs you take and the people you hire. Of course, you'll use more subcontractors than you would have as a capital-investment contractor. But that, too, has its advantages. There's always someone to "hang" when the plumbing, electrical, or drywall work isn't up to par. Someone else's insurance company will handle the claims. Someone else has to worry about collecting.

Paper contractors need little office space. Your den, or even the kitchen table, may be plenty. Accounting and utility costs will be minimal.

The whole idea behind paper contracting is to get control of your life and your business. It's hard to remember that the objective is to drain the swamp when you're up to your adenoids in alligators. The problems of a big construction company can take up so much time that you lose sight of your original goal: making a good living so you can enjoy life more.

Paper contracting gives time back to you and your family. Workload is reduced by 50% or more. Of course, gross income is lower. But that's no problem if obligations drop proportionately. And that's the other half of the paper-contracting adjustment. There's no way to keep the current staff and reduce your job load. Cut overhead, staff, equipment, and interest expense more than the workload falls and you'll thrive as a construction contractor.

In an earlier chapter I suggested that the more work a company does, the smaller the profit margin in each job. Let's look at the other side of the coin. It's time to do less work and make more profit per job. See Figure 7-2.

Remember, if you build only four or five jobs a year, there are only four or five owners to deal with. Each client gets more of your time. When you build less, you can do more of the actual supervision yourself. That cuts the payroll and reduces the number of subcontractors. And, by doing more of your own supervision, you should eliminate at least a few of the errors that cost you thousands of dollars in the past.

All in all, working as a paper contractor offers many benefits. About the only thing you have to lose is your problems.

Doing Your Own Work

Look, I know you're tired. It's all that worry and effort you've been putting into "fire fighting." Odds are, you spend most of your time repairing yesterday's mistakes, covering your checkbook, stalling the bank and your lenders, or meeting with your attorney.

Actually, construction work is only half as tiring as the mental strain of financial problems. The trick to surviving in the construction business is to simplify your life. You do this by reducing the number of potential problem areas. And remember — there's nothing wrong with doing your own work with your own tools, and skinning your own knuckles in the process.

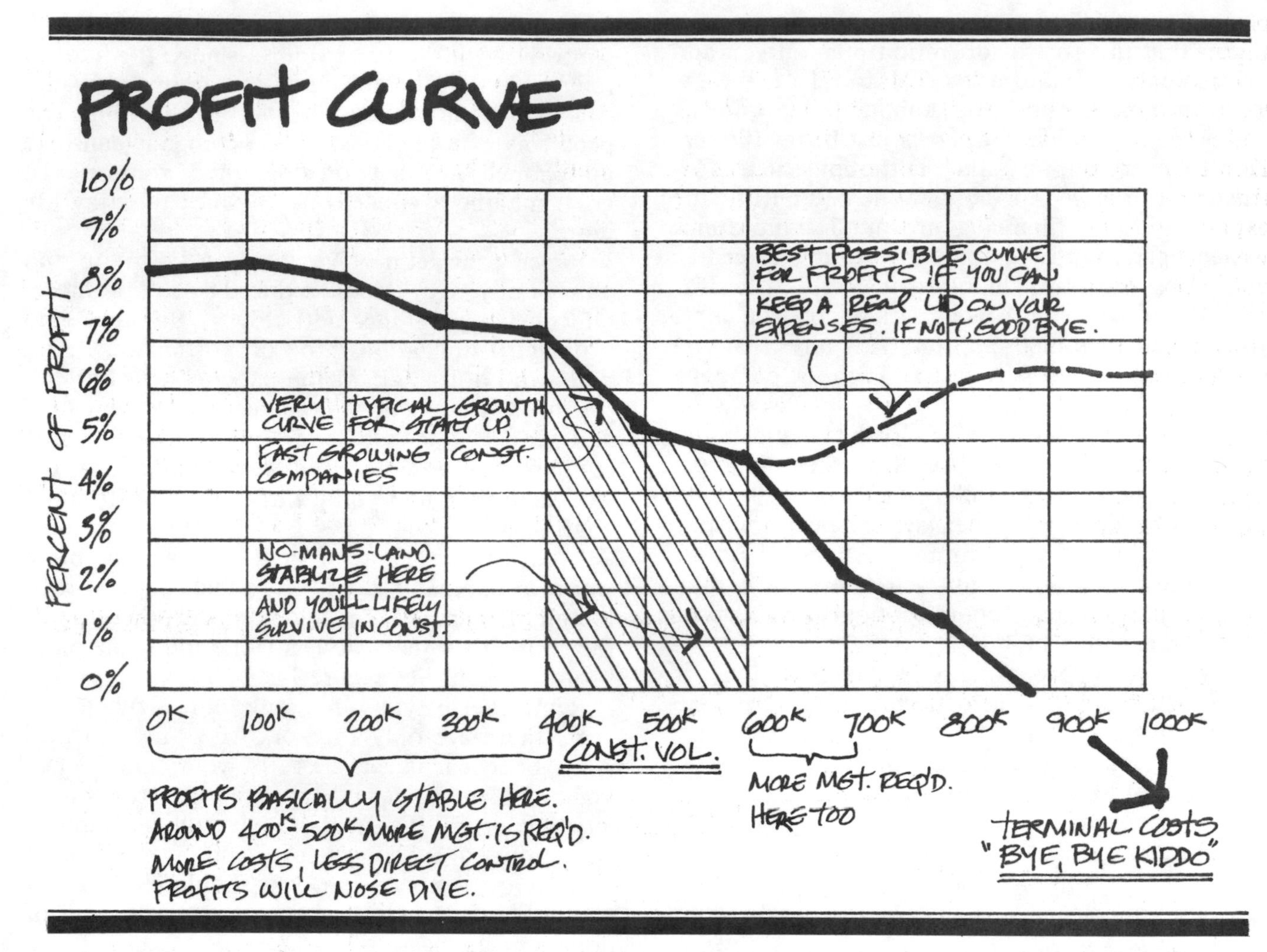

Profit curve
Figure 7-2

There's something therapeutic about doing your own work. I suspect that it comes from the control you have over your destiny. When it's your own hammer that's doing the work, you know at the end of each day what's right and what isn't.

Here's another benefit. Working on your own jobs is a good place to begin building the organization, if that's what suits you. Staying small enough to work on your jobs gives you both control and time to think about what you're doing and where you're going. It also gives the time you need to get back to solvency. Once you've established a solid base of work and a steady income, then think about expanding.

One final advantage to working on your own jobs: When you cut your own materials, there's sure to be less waste. Beams and joists cut too short, the wrong kind of wire, bad taping jobs, and a few hundred other mistakes won't happen any more because you're there. Theft should be lower too.

Equipment Maintenance

Taking care of equipment costs big bucks, no matter how much or how little equipment you have. It's tough to do even routine maintenance when you're up to your wallet in poverty.

The cost of equipment isn't just debt service. Equipment cost also includes insurance, interest, late charges, fuel, oil, air filters, maintenance parts, the equipment operator, and taxes. See Figure 7-3.

For some of the projected costs in Figure 7-3, the loan payment is only the tip of the iceberg. What will kill you lies hidden just below the waterline. Don't be one of those builders who succumbs to the temptation of owning lots of equipment, then expands beyond his ability to finance and control what he's got. The question at this point is whether you'll become a statistic or a survivor.

How Many Jobs in How Many Hours

It's Thursday, 6:00 A.M. Your foreman calls. He needs lumber. The lumber can't be picked up, because you're 60 days past due on your bill. Earlier, your other foreman called. He needs sheet metal and electrical supplies. Your foreman can't sign on your account at the electrical supply house, so you'll have to do it yourself. But nobody's open for business yet.

The phone rings. It's the tile setter. Mrs. Andrews is upset because the $600 tile job you installed doesn't match the sample you showed her. She wants you to fix it, right away.

You have three jobs waiting for a bid. Two of them close tomorrow afternoon. You can't find your plumber, to get a price from him. Your painter and roofer won't even talk to you, because they haven't been paid for the last three jobs they did for you.

You go through your messages from the past few days. There are six calls from Mr. Harper. He's more than a little upset. Nothing's happening on his job. Your framer was only there three days last week, he says. If he doesn't see some progress by Friday, he's going to call his attorney.

EQUIPMENT COSTS

NO.	ITEM	MONTHLY PAYMENT	INSUR.	MAINT.	INTEREST ON LOAN	DOWN TIME
1.	TRACTOR & TRAILER	#1200^{00}	# 65^{00}	#275^{00}	#175^{00}	35HRS.MO.
2.	BACKHOE	1650^{00}	110^{00}	325^{00}	250^{00}	18HRS.MO.
3.	DUMP-TRUCK	950^{00}	55^{00}	250^{00}	150^{00}	12HRS.MO.
4.	BOBCAT	375^{00}	45^{00}	125^{00}	75^{00}	27HRS.MO.
5.	FORKLIFT	1050^{00}	120^{00}	275^{00}	160^{00}	22HRS.MO.
6.	PICKUP	275^{00}	25^{00}	110^{00}	60^{00}	8HRS.MO.
7.	TOTALS	#4460^{00}	# 420^{00}	#1360^{00}	# 870^{00}	122HRS.MO.

PLEASE NOTE THAT YOUR TOTAL EQUIPMENT COST IS NOT #4460 BUT RATHER THAT COST PLUS #420^{00} INSUR., PLUS #1360^{00} MAINT., PLUS #870^{00} INTEREST, PLUS 122 HRS OF DOWN TIME AT #40^{00} PER HR. OR A GRAND TOTAL OF #11,990^{00} EACH AND EVERY MONTH WITHOUT FAIL AND THAT DOESN'T CONTEND WITH THE LOSS OF PROFIT THAT MONEY COULD HAVE EARNED HAD YOU INVESTED IT IN REAL ESTATE INSTEAD OF EQUIPMENT.

Equipment costs
Figure 7-3

You look at the message again. You were hoping to collect from Mr. Harper today, because you've got $5,500 in payroll due tomorrow. So far, all you've collected is $850. Your wife told you last night that your bank account is overdrawn by $1,925. She also mentioned that the tractor payment is 45 days late.

As you head out the door to drop the kids off at school, you remember that the phone's on a "24 Hour Shut-off Notice." You owe the phone company $287. Try to remain calm! After saying goodbye to the children, you head for the accountant's office. Your bonding agent wants an updated financial statement and a work status report for the bids tomorrow. As soon as you've pacified him, you have to take a tax deposit down to the bank. The IRS called Tuesday. They said if you don't make the payment by 5:00 P.M. today, they'll lock up your equipment.

Does any of this sound familiar? It's no wonder you're going nuts. Anybody would with this load of worries. It's time for a breather. But where are you going to start?

How about taking fewer jobs? Sound a little crazy? Maybe it is, but there's a limit to the number of jobs you can handle. If you're making money handling six or seven jobs a year, loading up with more is only asking for trouble.

Jobs demand attention, financing, and organization. The question is, how far can you can stretch your staff, your money, and your skills? Depending on the sizes and locations of the jobs, one foreman can only handle two projects — at best. Now, how many foremen do you have and how long does it take you to complete a job? Three months? Six months?

Let's look at the number of jobs you do this way. We'll say you have two foremen. Each can run two jobs apiece. Each job takes roughly six months to complete. Figure it out. The maximum number of jobs you can undertake successfully is eight. This assumes that you do no actual construction supervision work yourself. Your duties as owner are to handle the accounting, finances, taxes, materials, insurance, and of course, the clients. If you're not taking care of these aspects of your business, who is?

The only way to build more than eight projects a year is to extend your working hours. That means extra payroll, taxes, and insurance. It also means that you won't be home as much as you should be; this will undoubtedly take its toll on your family life. You'll pay the penalty in your children's bad grades, lost vacation time, and an upset wife.

Here are some guidelines to help you decide how many jobs you should take on at one time:

1) Assume that each project superintendent will spend at least 25% of his time just keeping general records of what's going on with his projects. So, in a 40-hour week you'll lose about 10 hours to the general administration required by your company policies. If that's all you lose, you've got a pretty efficient superintendent.

2) If your superintendent is handling a project of $500,000 or more, assume he can handle no other work until the project he's on is at least 75% to 90% complete. And yes, project complexity is a factor here.

3) If you're doing small jobs, in the $10,000 to $50,000 range, assume that your superintendent can manage no more than three to four projects, at best. And that's only if they are close to each other — no more than 15 to 30 minutes apart.

4) Assume that you can do no actual on-site supervision yourself unless you're handling fewer than four projects a year, each of which has a dollar volume of $125,000 or less. You'll have to stick to sales and company management if your construction volume exceeds $500,000 and is composed of more than four jobs a year. To succeed, you'll either have to hire a good superintendent or take on a qualified construction partner.

5) Finally, keep in mind that, whatever volume you decide to take on, it all must be staffed, managed, controlled, and accounted for. If you don't, then you're a goner for sure.

Playing the Other Man's Game

Everyone knows that the grass is always greener on the other side of the fence. But a funny thing happens when you get to that greener grass. It still has to be mowed, just like yours.

I don't care what kind of construction work you do, there's always some guy who seems to be making more money doing another kind. I'd like to have a hundred dollars for every time I've heard a hotshot builder brag about how much money he's making. Let me give you a piece of advice. When some well-meaning loudmouth begins to lay his success story on you, ask him how much he has left each month when all the bills are paid. He may be doing $100,000 a month, but you can bet he ends up with $3,000 to $4,000 clear, on which he has yet to pay his taxes.

Once in a while, you'll run into a guy who really does have his act together. He's usually the quiet type. You'll have to force him to talk about his financial affairs. If he says he's making $10,000 on a $100,000 monthly gross volume, ask to see his books. If it's true, drop what you're doing and follow him around like a puppy dog. Any guy making a 10% profit, year-in and year-out, is a genius. He deserves all the recognition he can get.

I'm pointing this out because many contractors tend to hop from one part of the business to another, searching for "the right spot." I've got some bad news for these guys. They'll grow old before they find it. The right spot is any spot that makes something approaching 10% clear profit on the volume, year-in and year-out. But finding it has more to do with the contractor who's running the show than with the kind of work done.

A simple description of a successful construction business is a 10%-profit business, the right size, financed right, and run right.

Remember, too, that switching emphasis can be costly. If you're in the concrete business and decide to take up framing, you'll need different tools.

Maybe you're building medical or office buildings and hear stories about profits in restaurant construction. How long will it take you to figure out the pricing and scheduling, and to develop a list of reliable subs who handle restaurant specialties? Do you have the cash to support yourself while you learn all the angles? Besides, what are you going to do with the equipment you have now? Sell it for 25 cents on the dollar?

Switching emphasis isn't a very practical idea for most builders. Stick to what you know best.

Figure 7-4 shows why a one-man operation will usually make the most money for the effort expended. Of course, it will also take more of your time. When you hire your first few employees, the profit percentage will usually drop. Hiring employees tests your ability as a manager, not as a builder. If your management skills are adequate, if you watch your manhours, cut waste, avoid mistakes, work well with people, keep good books, pay taxes on time, develop a good insurance program, and stay on top of collections, you have an excellent chance of avoiding a headlong crash into the zero-profit lane. Otherwise, you're just running an expensive hobby.

Profit percentage is the best measure of success in your business. Let's say that you hire a few employees and find a way to pull profits out of their initial crash dive. Profits start to creep back up like a car accelerating — past 3%, 4%, and 5% to maybe 10% or more. Someone in your organization is watching the profit speedometer. That person has to avoid mistakes, disputes, thefts, delays, strikes, lost time between jobs, sick employees, lawsuits, bad investments, and everything else that can go wrong. They have to react to the outside pressures that every builder has to deal with: high interest rates, material shortages, bad weather, high land prices, low buyer interest, and governmental regulation. Profits are extremely sensitive to both internal and external pressures. But profits are the best measure of progress in your construction company.

Growth and profits are not the same. Don't think of a big staff, a plush office, and loads of equipment as indicators of success. Growing is easy. It's paying for growth that's tough. Nothing will shred your cash flow like over-expansion.

Look again at Figure 7-4. Notice how the profit curve drops as you add people. Most new employees are not worth their pay until they learn the ropes and can pull their own weight. Expanding too fast puts too much deadwood on your payroll.

Here's how to find the right level of business in your operation. Constantly monitor your profit after all expenses, including your salary. As time passes, your profit margin will rise and fall. The question is whether the change is due to fluctuations in the economy or to decisions by management. Of course, this is really a judgement call on your part. But making this evaluation is important to your survival as a builder. If profits are dropping because you've expanded beyond your ability to control the company, relax. That's normal. It's nothing to be ashamed of. Finish up whatever's draining your profits, be more selective about taking new work, and cut staff back to where your profits begin to rise again. It could take 6 to 12 months to get back to profitable operation. You'll have enough time to plan your moves, provided

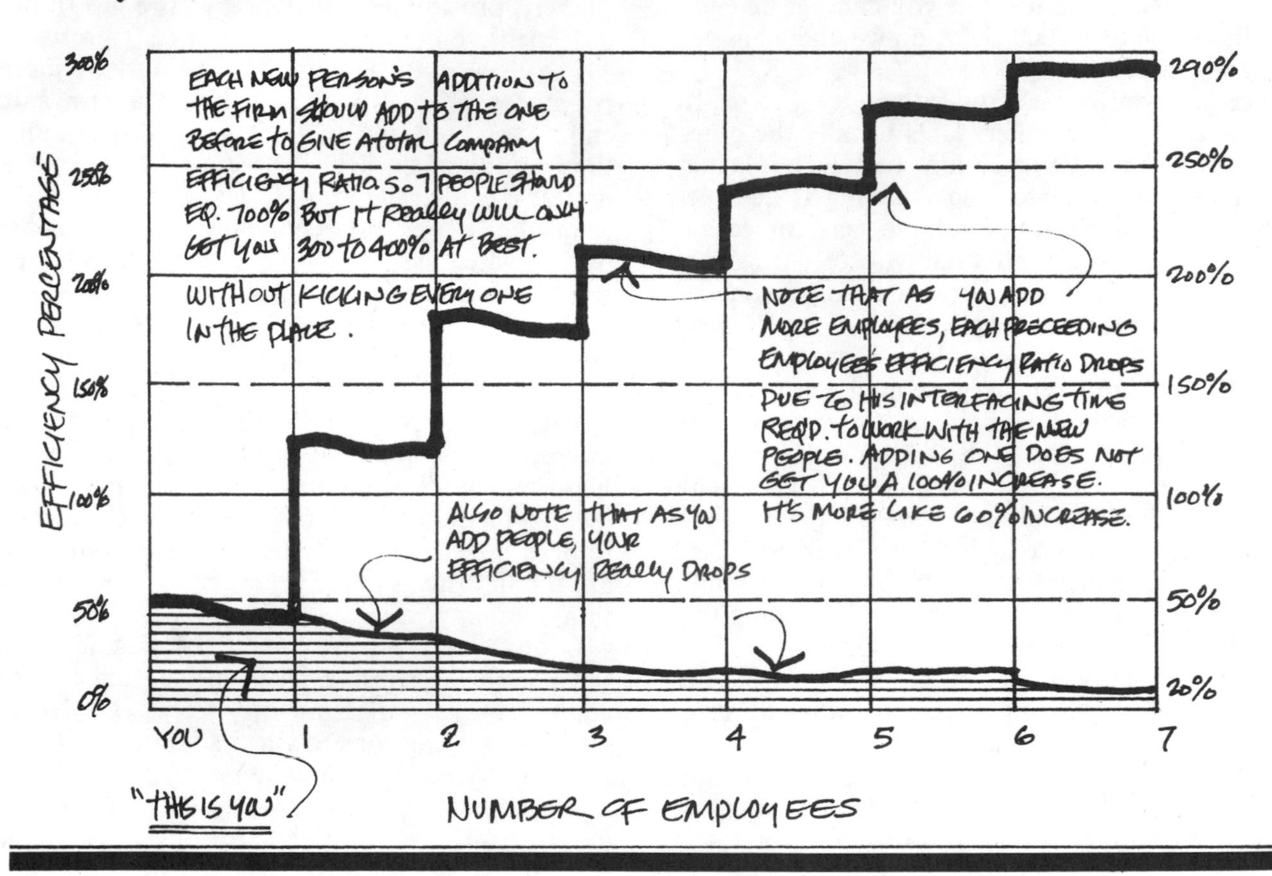

Employee efficiency and you
Figure 7-4

you aren't up to your neck in upset owners and legal problems.

Your Working Radius

Every builder has a comfortable working radius — the distance between his home base and his most remote job site. Time spent getting to and from a job can be a killer. Work done out of town may not be as profitable as work done in town at the same price. Figure 7-5 shows a typical working radius.

Everybody builds within 30 minutes of home. Transportation and lost time are very minor costs in that area. But longer drives are tough on people, equipment, and profits. Workers will hire out for a couple of dollars an hour less just to stay close to home. You may have to pay a premium to get tradesmen, suppliers, and subs to work remote jobs.

You can probably cope with projects within a 60-minute drive, although the costs will be higher. Unless the project is profitable enough for you to move in staff to live there until it's completed, avoid any jobs over an hour away.

In remote jobs, the biggest loss is your own time. Figure the time required for each round trip. Then multiply by the number of trips expected. Figure your cost of driving at $25 to $50 per hour and you'll see the cost of work beyond a reasonable distance.

WORKING RADIUS

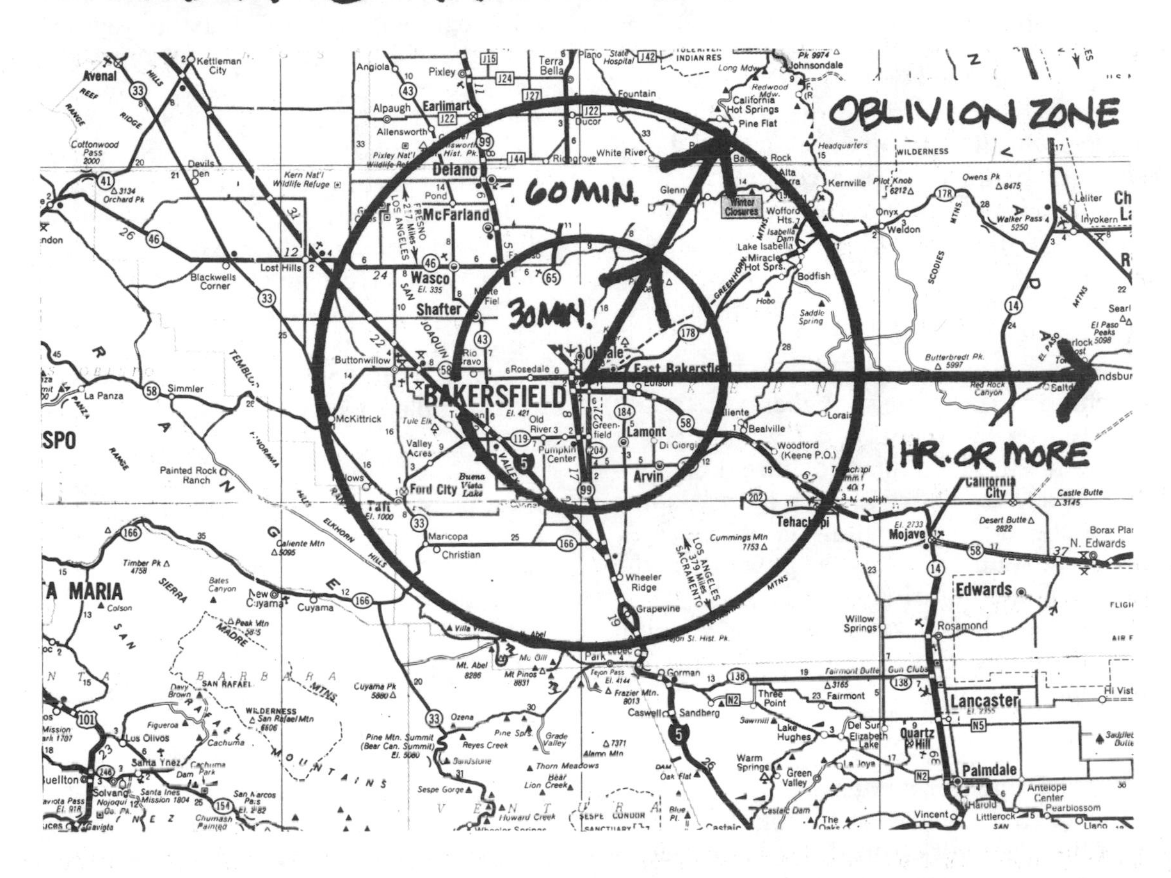

PROJECTS WITHIN 60MIN. DRIVE EACH WAY CAN BE COPED WITH, WITH SOME EXCESS COST. BUT UNLESS THE PROJECT CAN AFFORD MOVED-IN STAFF, ANYTHING BEYOND 60MIN. USUALLY BECOME DISASTERS AT BEST. AVOID THE PROBLEM & STAY HOME!

Working radius
Figure 7-5

Of course, added driving time isn't the only extra cost. Higher phone bills, gas bills, maintenance, taxes, insurance premiums, and city licenses are likely for remote jobs. How are you going to stay competitive with that burden around your neck? And how can you look after local jobs when you're out of town? It's something to think about.

So You're a Lousy Salesman

Contractors must be the world's worst salesmen. Salesmanship is a fine art. It takes lots and lots of patience, and most of the builders I know have very little patience. Salesmanship takes study, vigilance, and practice. It takes know-how, inside information, packaging, and familiarity with your client and his needs. Big, solvent corporations give jobs to contractors who are consistent, helpful, and available at all times.

Who's going to do your selling? Do I even have to answer? You are, of course. It's up to you. No one is going to knock down a door to drop a job in your pocket.

You'll need references. I don't mean a couple of fishing buddies who'll say that you're a good guy. I mean respected members of the community who are your clients. Hopefully, these clients will vouch for your honesty, sincerity, diligence, pricing, and ability to meet a schedule. Be ready to provide the names of banking and insurance connections to potential clients.

Finally, prepare a handout or brochure that you can leave behind with a potential client. It should list your references and mention a few major jobs that you've handled. Have this brochure done by a professional artist. Don't try to save a few dollars here. The brochure is your calling card. It's probably more important for sales purposes than the car you drive, or the tailoring of your suit.

Don't continue to be a lousy salesman. Wake up and do a little careful planning on your sales efforts. It will make a world of difference in the jobs you get.

Repeat Business

The easiest place to get new work is from the clients you've already done work for. There's no better referral than a job well done. It takes very little effort to get a second job from a client who likes what you did the first time around.

Here's another good thing about repeat business. Your competition has an uphill battle to separate you and an established customer. Your only sales problem is pricing, not fighting fifteen other contractors for a client's trust and attention.

Keep your business in decent shape for five or six years, satisfy clients for that long, and about half of your work will be for repeat customers. A sound construction business develops a reputation that sells better than thousands of dollars in advertising. The public knows that there are too many suede-shoe operators in construction. Every owner wants to avoid cheap-jack contractors. A builder's good reputation separates him from the herd of slipshod fast-buck artists.

Repeat business is the best kind of business you can get. It increases your profit margin because there's little sales cost and no advertising costs. There will be fewer meetings with the client because you begin with a solid, trusting relationship.

There will be little repeat business in the first two or three years of your business. But if you've done your job right, you'll still get some nice referrals from satisfied clients. Eventually you'll sow enough seeds to keep the harvest perpetual without more cultivation. An owner who has built once is likely to build again. Keeping clients satisfied reduces the effort needed for sales.

Do yourself a favor. Cultivate good relations with your clients. Cultivate clients who build frequently. You'll profit by the association. You'll also be able to survive hard times.

Summary

If you plan to keep your sanity as a builder, find a profitable niche for yourself. Find a type of work that fits your skills, personality, and lifestyle. Bigger isn't necessarily better. Capital-intensive contracting can take a lot out of you, perhaps more than it's worth. Paper contracting is an excellent option. You can build fewer jobs, cut back on payroll, sub most of the work out, and do a lot of the actual supervision yourself. The result can be a decent profit and fewer headaches.

Don't overlook the option of working along with the tradesmen on your jobs. Usually the only contractors driving nails are new builders and builders in financial trouble. For them it makes sense. It's the cheapest and most productive way to get the job done. Maybe it's the best way for you too.

If profits are dropping because you've overextended, finish current jobs and then cut back on staff until you find your optimum level of operation. Make yours a successful construction business: 10% profit, the right size, financed right, and run right.

Take time to get your sales package together. Include a brochure you can leave with potential clients. The best referral, of course, is a job well done. Cultivate clients who need to build repeatedly as they expand their businesses. If they like what you did the first time around, they'll be back for more.

8 One Problem at a Time

When you take on a new client, you also take on his wife, relatives, and business associates — for better or for worse. The family relationship isn't limited to immediate family members, but extends to the business family of your client as well. After his wife, his secretary, his office manager, or his bookkeeper may be next in the chain of importance to you as a builder. Don't assume that these people can be overlooked. To succeed in construction contracting, you have to be aware of these relationships and sensitive to the wishes of others.

Here's an example of what I mean. Let's say you're building an addition to someone's house. Your drywaller's a slob. His crew tracks up the carpeting and tosses lighted cigarettes at random as they work. Chances are you'll be hearing from your client's wife very suddenly.

Here's another example. You're doing a small fill-in job between two larger jobs. Of course, the owner doesn't see his work as either small or fill-in. Still, everyone's happy until the five-man crew doesn't show up one day because it's needed on the next job. From your standpoint, this smaller job is like moonlighting. It takes a back seat to the real action. But imagine your client's reaction when he gets home from work and finds that nothing has happened all day. You're sure to get a call from one rather upset client — all because of a lack of communication and common understanding. You knew what you were up to, but he didn't. You were asking for trouble by not spelling things out for him.

Keeping wives and associates happy doesn't have to run up your expense account for candy and flowers. It only requires that you act like a professional, and do just about what everyone expects you to do. Failing to connect a stove, or shutting the electricity off before checking with the occupants, will only guarantee that you'll soon hear about your shortcomings. Never discount the importance of your client's wife or business associates. Either could be signing your paycheck.

Your Home, Your Home Phone and Your Privacy

Your family deserves a quiet, peaceful, protected place to live. Running a business out of a living room may be bad for both business and family. Your family members get involved in the business just by answering the telephone. That's the first point of contact with new clients. An angry client needs more personal treatment than your five-year-old can provide. When your plumber's upset because he hasn't been paid, who gets to explain to him that you're not in and won't be back for several hours? When your checks bounce, is it your wife who gets to explain why? These are the penalties of having a business phone at home.

Business is business, and your family life is too precious and delicate to let it be upset by problems associated with construction. Your home is your entire family's castle, not just yours. Your children, bless their hearts, couldn't care less what you do for a living. And why should they care? Children have other problems to deal with — like who ate the last of the peanut butter, and how to get out of doing the dishes. Naturally, you'll have your kids out on the job site sometimes to clean up and to do minor work, but their involvement shouldn't be continuous and involuntary.

My advice? Simple. Separate business life from home life. Give your family a break. Get the business office and phone out of your home, even if it's only to the garage.

Time Off

Time off, if you don't already know it, is probably your most precious reward. Once your company is making money, the idea is to get more free time. To do this, you'll need to police the hours you work carefully. Six A.M. to six P.M. is plenty. You're only 50% efficient after that, anyway. But make your work time count. Work on jobs that pay. Work hard when you work. But when the day is over, quit and go home. You deserve any time off you can get.

If you can't afford two weeks in Maui, how about a couple of four-day weekends camping, fishing, and hiking in the mountains or at the lake? If money's tight and you and your wife are tired of the kids, try hiding out at a friend's house while he's out of town for the weekend. Two days off, a nice dinner, a bottle of good wine and some soft music can do wonders for your married life.

Employees

Select employees and associates for their intelligence, skill, appearance, and the capability they add to your company. Some subtle and not-so-subtle points deserve consideration: how long they've been married, if they rent or own their home, how long they've lived in the area, if they smoke or drink, who they've worked for in the past, and if they have a criminal record. It's hard to determine from a short interview whether someone can think for himself. But that's important in many jobs.

Will an applicant be able to keep you and your project out of trouble? Remember, the people you hire are your first line of defense against legal distress. Look at it this way. If you don't have the time to find the right person for the job, you don't have the time to pick up the pieces when someone louses up your project.

Summary of Part 1: Surviving

We've covered quite a bit of ground so far, from the first chapters when we discussed kiting loans and stalling creditors; to this chapter where we started planning to hire new employees and to take more time off. What you learned when survival was in doubt should be the foundation of your success as a construction contractor. Don't forget those lessons as you read the second part of this book. Every sound and profitable construction company needs a firm foundation.

The remainder of this chapter summarizes the points that make up the foundation of your successful construction business.

Every contractor struggling to survive needs accurate information on his financial position. When you're in trouble, draw up a financial statement before trying to satisfy any debts. You're the only one who knows the amount of each debt and the value of each asset. If you don't compare debts and assets, you'll never know exactly where you stand. Builders who don't have a current balance sheet get derailed by even a minor hiccup in their cash flow.

Don't dump any hard-earned cash into your debts until your financial statement is in order. Bankruptcy may be the only way out. But a little planning and some cash could bring you back to solvency.

As you recover, use the inflation-recession cycle to your advantage. Avoid the devastating effects of the inevitable economic downturn. Be prepared for the equally inevitable recovery. The only differences between victims and opportunists in the construction cycle are planning and timing. Every downturn has a certain depth and a duration. Learn to anticipate these swings.

In Chapter 2, I advised you to rate each debt by priority. Some are more important to your survival than others. But every debt has a place on your priority list. Setting priorities merely puts you in control.

Once debts are listed in categories, according to their ability to bite you, allocate income and assets to each. Some holdings are more liquid than others and can be used more easily to retire debt. But all obligations can be met, given enough time.

Assets that can't be converted to cash easily can still be used to reduce debts. Frozen assets can be swapped for debt or can serve as equity for new loans.

A debt-reduction program might include new borrowing. Sometimes new loans do make sense. But that only transfers the debt and postpones payment. That's not the real answer. Besides, once you've got a bank on the hook, you can do a lot of stalling and leveraging. The only real way out is to increase earnings. Concentrate on reducing costs and increasing earnings, not borrowing. But avoid giving a lender new collateral if possible.

Buy time from creditors with promises, trust deeds, auto pink slips, or stock. There are a thousand ways to stall for time. In fact, your creditors can probably suggest ways you haven't thought of. But staying in touch is the simplest method I know of. Moreover, it's the cheapest.

Higher earnings, income conservation, and debt avoidance save construction companies in financial trouble. Nothing else works. To survive, you have to quit spending and work exclusively on what's going to yield immediate income.

A good tax man is essential to your debt-reduction program. It's possible to owe taxes even though you didn't receive any cash. Get a competent accountant — it really helps. And have three or four checking accounts. Let the bank help keep your accounts straight. And, using several accounts (sparingly) can buy much-needed time.

Sometimes, time and money aren't enough to solve your problems. There's just too much debt. Chapter 4 explains bankruptcy procedure. Don't go the bankruptcy route without counsel from a bankruptcy attorney. If you avoid bankruptcy, you'll be facing lawsuits and bills. For this you'll also need a qualified attorney.

Lawsuits have a way of going on and on. Believe it or not, that works in your favor when you're broke. Any debt that's in dispute can drag through the court for years. Time lost rarely injures the debtor. And the creditor's memory, resources, and finances weaken with time. You could get lucky if enough time passes.

Fighting your way out of debt may require transfer of assets. That may avoid premature liquidation. Use trusts, corporate forms, friends and relatives to hide your assets. It may be to your advantage to become technically broke as soon as possible. A builder who's broke is a protected builder, one who sleeps easier at night. It's up to you to protect yourself. Nobody's going to do it for you.

If it isn't already, get your bookkeeping in order. When you're in financial trouble, sloppy bookkeeping only complicates matters. Do your accounting faithfully.

Look carefully at your equipment cost. How much do you pay each month? Do you use it full-time? If you don't, it could break you. Unload anything you're not using constantly. Rent it back when you really need it.

One more note before we head into the wonderful world of Thriving Builders. Do you have partners? Maybe too many partners? It could be time to lighten the ship. It's harder for a partnership to make decisions. You may go broke before the right decisions are made. If you're going broke gradually, it's easier to maneuver with less baggage. Here's a little gem of wisdom a man in Seattle, Washington, gave me years ago. "Don't take on a partner to do what you can hire done."

Part Two:

THRIVING

- Your Purpose and Goals
- Going Where the Money Is
- So You Can't Find a Job?
- Over-Design, Under-Design & No Design
- Second, for the Third Time
- Investing in Inflation

Your Purpose and Goals

To thrive in construction, you need to establish some goals. Otherwise, you'll wander from job to job, handling whatever work is available without focusing on the right opportunities.

What's your purpose? What are your goals? What exactly do you want from your efforts? What do you expect from the building business besides calluses? You probably work as a builder because it's something you're good at. You enjoy working outdoors and with your hands. And there's real satisfaction in putting up useful, durable, attractive buildings. But you also want a reasonable reward for your effort.

Let me give you a simple lesson in goal-setting that I learned from an insurance man turned real estate developer. When I first met him, this man had just sold his insurance business. With the proceeds he went into the investment business with several partners. The first year, they made only 13% on their investment. That simply wasn't enough to live on. So, he left his partners and began investing in real estate. In the next year he more than doubled his money. He told me it didn't take him very long to figure out where he wanted to spend the rest of his life. Several years later I ran into him again. This time he was a millionaire. The secret I learned from him was the setting of goals. He taught me the method I've used ever since. It's really quite simple, yet it's very effective.

Begin each year by preparing a personal financial statement, just as explained in Chapter 2. This statement is your reference point for the year. At the beginning of the next year, draw up another financial statement. Compare the changes for the 12 months. How much was gained or lost? What's the change in your net worth? Are you winning, losing, or just breaking even?

When making up that financial statement, make a list of goals for the coming year. The goals should be reasonably attainable within the next 12 months and should cover three areas:

1) Personal goals
2) Professional goals
3) Financial goals

Under each heading, set three major goals for yourself for the coming year. Figure 9-1 shows some of my own typical goals.

Don't be too rigid in your thinking. Leave a little room for changing or amending each goal as the year goes on. That's to be expected.

GOAL SETTING (FOR 12 MONTHS ONLY)

1. PERSONAL GOALS:
   A. SPEND MORE TIME WITH THE FAMILY (3 WEEKS VACATION).
   B. LEARN TO PLAY THE GUITAR (SIGN UP FOR LESSONS & SHOW UP).
   C. PLANT A VEGETABLE GARDEN WITH MY KIDS.
2. PROFESSIONAL GOALS:
   A. TAKE A COURSE IN BUSINESS & ACCOUNTING (SIGN UP).
   B. INCREASE BUSINESS VOLUME BY 15% (MORE SALES TIME).
   C. LOCATE A BUSINESS MANAGER FOR PARTNERSHIP (NOT ME).
3. FINANCIAL GOALS:
   A. INCREASE SAVINGS ACCT. BY $5,000.00
   B. INCREASE BUSINESS CASH RESERVES TO $20,000.00 (NOW $10K).
   C. CUT ALL PERSONAL & BUSINESS EXPENSES TO THE BONE (NOW).

Goal setting
Figure 9-1

GOAL IMPLEMENTATION

1. PERSONAL GOALS:
   A. SPEND MORE TIME WITH THE FAMILY (3 WEEKS VACATION).
      1. DECIDE WHEN & ALLOT TIME ON CALENDAR.
      2. DECIDE WHERE & MAKE NECESSARY RESERVATIONS.
      3. SET ASIDE THE MONEY REQ'D.-$175.00 MO. (DON'T TOUCH IT).
   B. LEARN TO PLAY THE GUITAR (SIGN UP & SHOW UP).
      1. FIND AN INSTRUCTOR AND SIGN UP NOW.
      2. BLOCK OUT THAT EVENING ON YOUR SCHEDULE.
      3. BUDGET THE MONEY SO YOU REALLY DO IT ($30.00 MO.).
   C. PLANT A VEGETABLE GARDEN WITH MY KIDS.
      1. MAKE THE PLANS (TELL THE KIDS) GET ORGANIZED.
      2. DECIDE WHERE TO PLANT IT. (LET THE KIDS DECIDE).
      3. GET THE STUFF (FERTILIZER, SEED ETC.) & GET GOING.
2. PROFESSIONAL GOALS:
   NOW DO EXACTLY THE SAME TYPE OF LIST HERE AS YOU DID ABOVE, FOR THE REST OF THE ITEMS ON YOUR GOAL SETTING (FIGURE 9-1) OUTLINE AND YOU'RE READY TO START.

Goal implementation
Figure 9-2

When you've set your goals for the year, put the title "Goal Implementation" on a second piece of paper. On this sheet, write down each individual goal listed below the major goal headings. Below each individual goal, list the three or four steps needed to reach that goal. Mine is Figure 9-2.

When your list is complete, pin it up on the wall where you'll see it once a day. Mark it up, change it as needed, but work toward those goals every day. At the end of the year, check your progress. If you're like me, you'll reach 50% to 75% of these goals every year. I know the technique is naive. But the results aren't. Try it. What have you got to lose?

Committing goals to paper reinforces them in your mind. That leaves you free to consider the "how to's" rather than dwelling on the "what to's." Setting goals also helps crystallize your thinking. It gives you a clear and expressed plan of action that you didn't have before. *And it deprives you of excuses for doing anything less than the best you can.*

So much for annual goals. What about lifetime goals? Once you have annual goals nailed down, it's time to look a little further into the future. The place to start is with the major elements that make up your life. Let's look at *you*. What do you want out of life? What do you hope to accomplish? Look at your family, your business, your professional life, your finances, your assets, your plans for retirement. What we're going to do is fit all of this into a lifetime schedule.

First let's look at your age and what you've already accomplished. If you're sitting at mid-life, somewhere around 40-45 years old, then 50% of your time is gone. You'll likely want to eliminate some of the goals you had as a younger man, such as a new boat. A trip around the world might be more important to you now.

If you're at mid-life, don't discount raising your family as one of your life's major accomplishments. That's no easy task in anyone's book — and some people don't do a very good job of it.

Make a list of the things you would really like to see happen in your life. Not little things like a new job, but rather major elements such as paying off your house, a world-wide trip or a well-funded retirement account. Now take a sheet of paper and make several headings down the left of the sheet. They should be similar to the goals in Figure 9-1, only they'll be far more encompassing in their scope. The headings should include your assets, your family, your business, your retirement, recreation, and anything else that's important to you. Use headings that make sense to you, even though one or two may seem frivolous to someone else. If it's important to you, that's all that counts. Your list will look something like Figure 9-3.

Under the same three headings, personal, professional and financial, list the three or four most significant goals you have for each. Be realistic. Each should be attainable in your lifetime. I realize that choosing three or four lifetime goals per category is a tough proposition. You're likely to have many more than that. But go back through your list. Cull out a few choices that are more like milestone events than lifetime goals. Take out the goals that are actually short-term desires, such as an automobile, a boat, or a trip to Hawaii.

When you've reduced the list to a handful of key personal goals, it's time to find the cost of each. Some of your choices won't carry a price tag. But for those that do, estimate what the cost will be. See the "Money Required" column in Figure 9-3. You'll notice at the bottom of this column a total amount which I call the "Bottom Line." This is your estimate of the cost of meeting your lifetime goals. If the money doesn't meet the goal, reduce the cost of the goal or increase the earnings expected.

Next, compare the "Bottom Line" figure with your estimated lifetime earning capacity. Start with a small graph, which I call "Finance, Goal and Time Projections." Look at Figure 9-4. On this graph, plot your goals. Compare the goals and your income potential. Spread the cost of goals over the time required to do each.

The Finance, Goal and Time Projection chart is your roadmap to professional success. It puts your purpose in perspective and should provide a clear picture of where you're going with your life, your business and your finances.

A Realistic Look at Retirement

Many builders end up at age 65 ill-prepared to slow down and to enjoy the fruits of their labors. For that reason, I believe it's important to plan for your retirement from the very beginning of your productive career.

At best, your working years begin at age 18 and extend to age 70. This productive span is 52 years,

LIFETIME GOALS

GOAL DESCRIPTION	MONEY REQUIRED
1. ASSETS DESIRED	
A. PAID OFF PRIMARY RESIDENCE	$85,000
B. REAL ESTATE INVESTMENTS ($1,000,000)	200,000 DOWN PMS
C. CASH IN SAVINGS	
2. FAMILY DESIRES	
A. FORM A FAMILY PARTNERSHIP	-0-
B. SETTLE DIFFERENCES WITH THE KIDS	-0-
C. TAKE A MAJOR TRIP WITH MOM AND SISTER (COST SAME AS ANY TYPICAL VACATION)	-0-
3. BUSINESS DESIRES	
A. BECOME A SENIOR PARTNER IN A MAJOR CORP.	-0-
B. BE PART OF AN ASSET DEVELOPMENT CORP.	-0-
C. BUSINESS SELF-OPERATING BY AGE 60 SO I HAVE TIME OFF AT WILL.	-0-
4. FINANCIAL DESIRES	
A. ANNUAL INCOME TO BE $100,000 FROM ALL SOURCES	-0-
B. MEDICAL AND LIFE INSURANCE PAID IN FULL	25,000
C. NO OUTSTANDING BILLS OR OBLIGATIONS EXCEPT MORTGAGES.	135,000
5. RETIREMENT AND RECREATIONAL GOALS	
A. TAKE A TRIP AROUND THE WORLD	25,000
B. TAKE 6 MAJOR INTERNATIONAL VACATIONS	30,000
C. BUY AND PAY OFF OUTBOARD BOAT (SKI TYPE)	25,000
	$625,000

Lifetime goals
Figure 9-3

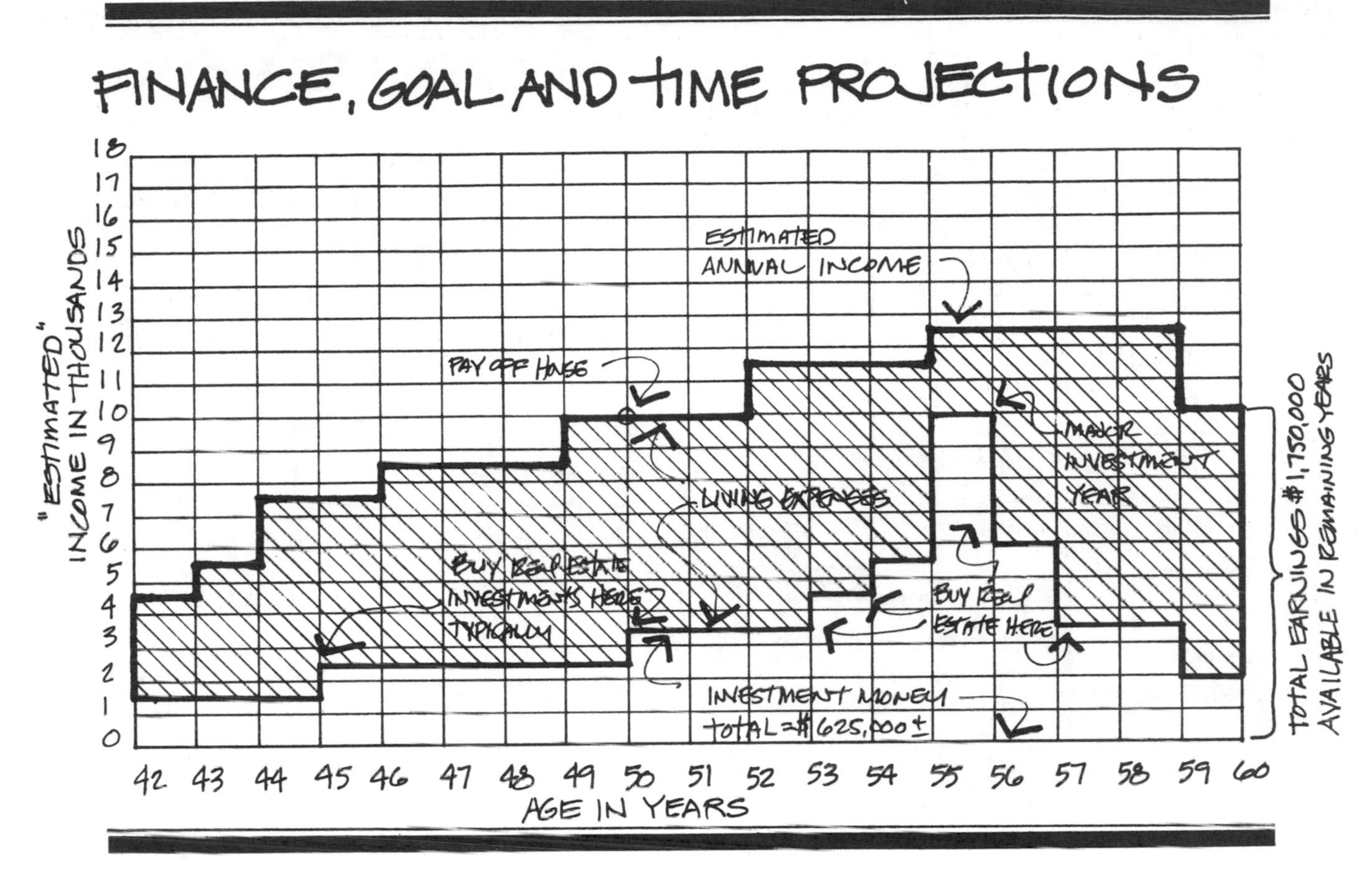

Finance, goal and time projections
Figure 9-4

the time available to earn whatever you need to retire. But you're most likely quite a bit older than 18. That gives you less time to meet retirement needs.

If you're never realistic about anything else in your life, be realistic about retirement. No one should plan to work forever. Part of being realistic is setting reasonable goals in five areas important areas. Work your retirement needs into a Finance, Goal and Time Projection chart, like the one in Figure 9-5. Now, let's itemize the requirements for retirement.

First, you need an income source. Income should come from rentals or retained assets, your "multipliers."

Next, you need reserves. These reserves aren't the same reserves your business has. If your business is to continue, it will need reserves of its own.

Retirement reserves are anything of real value that can be converted to cash if you become ill or incapacitated. It doesn't have to be cash. It could be any asset that is easily liquidated in time of need. In fact, converting to cash too soon is usually foolish. That ends the inflationary growth most non-cash assets enjoy. Generally, it's best to float with the inflationary bubble as long as possible. It's usually a mistake to keep your entire retirement reserve in cash in a bank, losing value faster than the interest earned compounds it. Instead, plan to liquidate only when the need arrives. The small penalty you may pay for a quick sale will be worth the cost.

Retirement years should be debt-free years. Don't plan to have any payments on cars, boats, furniture, college tuition and the like when you're retired. And the mortgage on your residence should be paid off if possible. That may seem like a

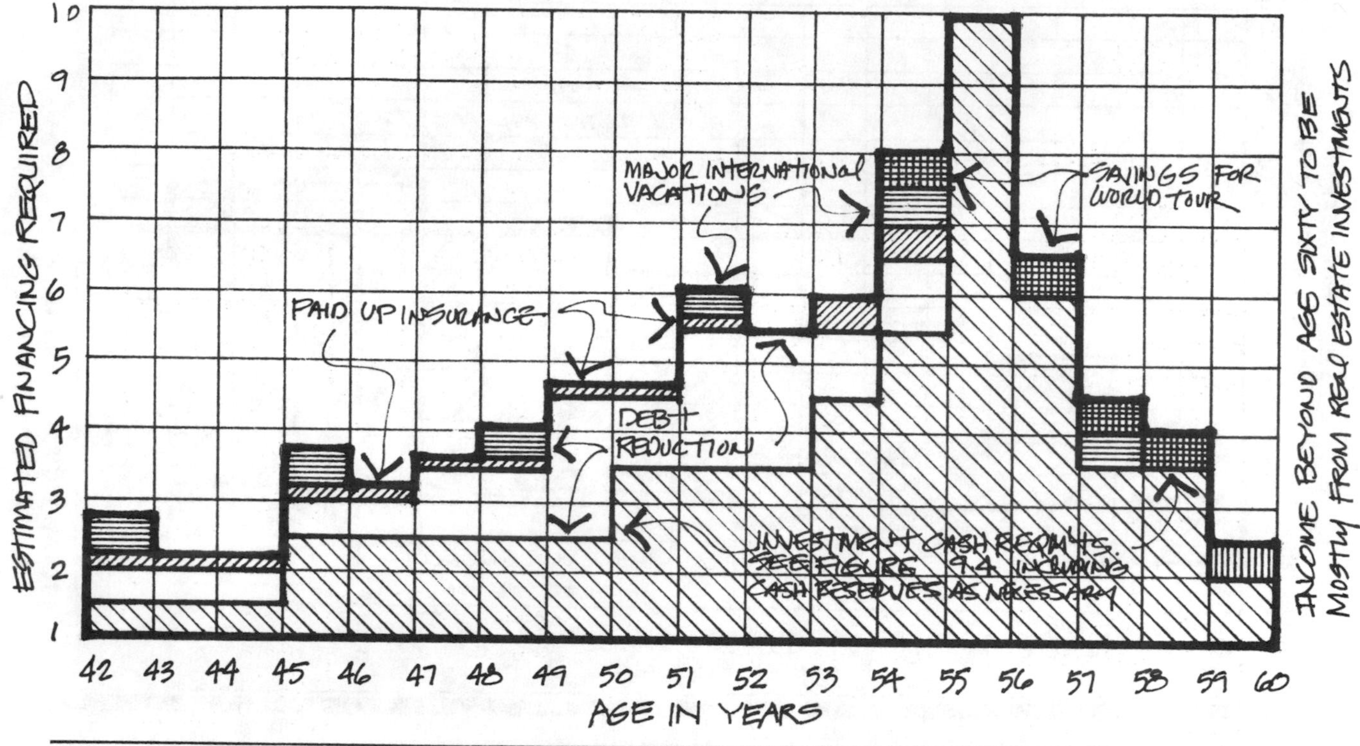

Retirement finance, goal and time projections
Figure 9-5

lot of money. But every successful builder should be able to handle it in his lifetime. Labor swaps and discounts can cut the building cost to about 30% below the going rate. An advantage like that should let you accelerate the payment schedule on the loan so the note is paid off at least five years before retirement.

Plan to have a good noncancelable health insurance plan in your retirement years. Life insurance isn't so important if you've managed to build up your assets. If you haven't, some term life may be the only choice, even though the cost is high at retirement age.

Be realistic about your future. Get organized today and project your lifetime plan. Start your own retirement program now. You'll be prepared when the time comes.

Your Work and Your Company

At least half of the construction firms operating today have very little, if any, direction. They bounce from one project to another until the owner either gives up because of financial problems or develops a good business plan with carefully selected and attainable goals.

What type of company are you developing? Form a clear idea of what you're building and where you're going in the construction business. Remember, the kind of business you build is not limited to just your present trade specialty, your size, your location or your working radius.

The most stable building businesses fall into one of two types. The first type is the small, closely-held, family-operated firm that's handed from father to son as each reaches his most productive

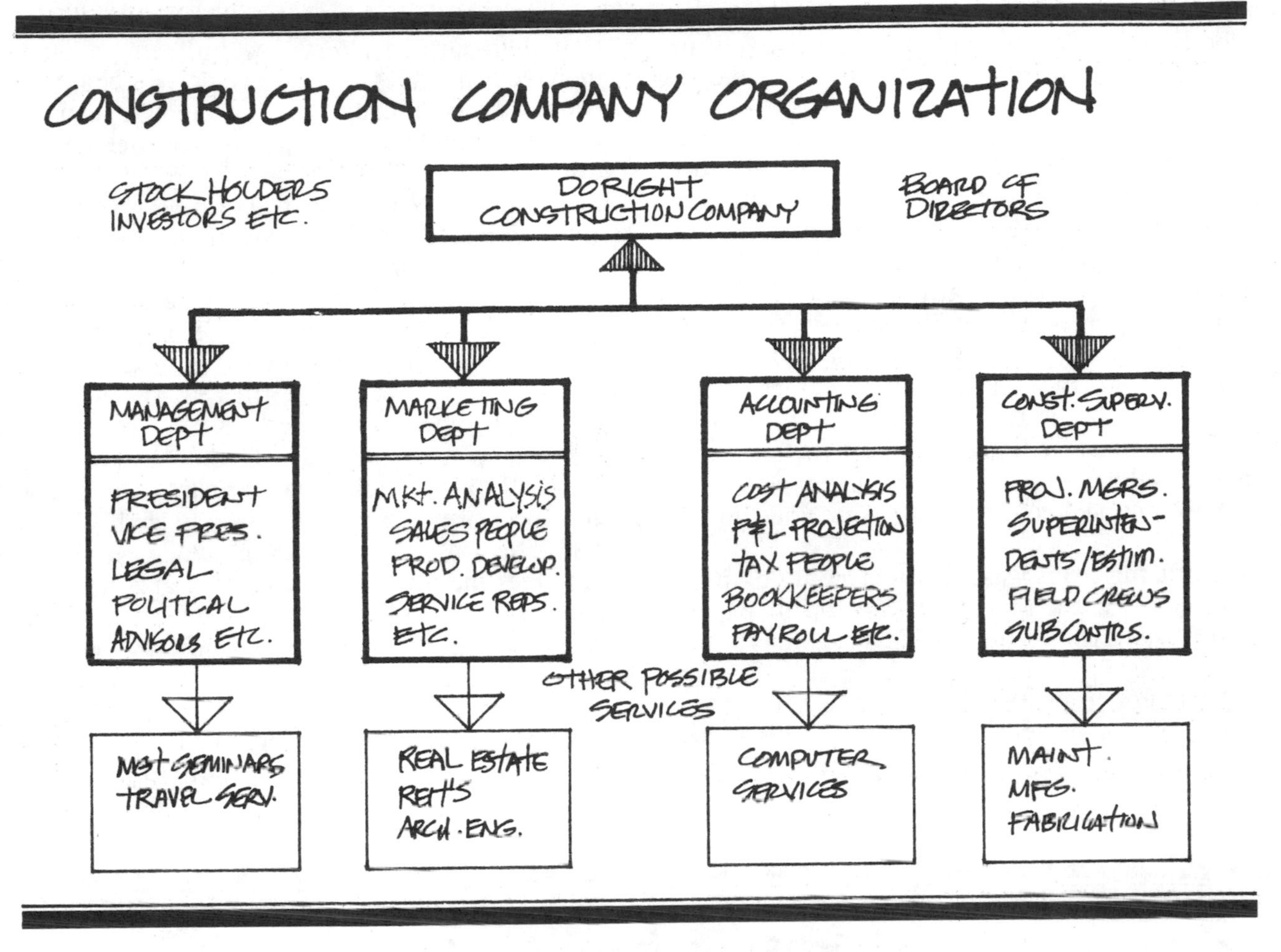

Construction company organization
Figure 9-6

years. Success is based on the skill and hard work of the owner. The second type is a larger corporation that has sold stock to the public. It depends on its financial muscle and hired managers to compete successfully in the construction market.

Decide which type is a model for your business. If you're like me, you'll decide to strive for a smaller, more profitable, closely-held company.

Next comes the question of your type of work or trade specialty. Are you a general contractor, framer, sheet rock hanger or fire-sprinkler specialist? Pick out your trade and stick to it. No contractor or subcontractor can compete in more than a very few specialized areas. Trying to run several trade specialties dilutes your time, energies and resources. Switching trades regularly to follow perceived opportunities is also a mistake.

In short, be selective in the type of work you do. Decide what you're best at. Which jobs will put the most money in your pocket? There's little point in handling general construction if what you really enjoy doing is concrete work. But if you can't show a profit pouring concrete, you'll soon become disillusioned with your efforts.

Once you've chosen your type of work carefully, marketing becomes essential. All contractors sell themselves. You have to learn to sell yourself and your company. Good organization, scheduling, purchasing, supervision and collection are irrelevant until there's a job to organize, schedule, purchase for, supervise and collect on.

Figure 9-6 shows that most construction companies have only four basic moving parts. That's why the most straightforward method of organization is the best. Complexity only tends to confuse everybody. There's no room for an unwieldy or cumbersome organizational scheme in a small construction company.

The first part of every small construction business is the owner. He's the originator, the guy with the drive, guts and ideas. Next comes the marketing arm. Somebody's got to sell. Nothing happens in any business until somebody sells something. The marketing arm of most small construction companies will be the owner. That's an additional task for the owner — in addition to responsibility for motivation and direction.

Next on the list is the construction supervisor. He's the "workhorse," the engine of the business. What you sell has to get built. Having experienced and honest people here can help ensure your future growth.

Last, but far from least, is the accounting puzzle. Somebody's got to be right on top of paperwork and cash flow if your business is going to survive.

One person can fill two or three or even four of these jobs in a company. But if the owner wears four hats, work seldom flows smoothly. Some tasks will be ignored while others get more attention than needed. Very few builders can handle management, sales, job supervision and paperwork equally well. Ideally, even a small construction business would have one person for each key function.

Of course, many construction companies, even small companies, have more than one owner. In fact, there are many ways to cut up the ownership pie. They vary from sole ownership to total employee-owned organizations. The most common form of ownership is the *key man* method: ownership is divided among the principles of the business in proportion to their contribution or seniority. For example, the founder of the company might own half of the stock. The marketing, construction and accounting managers would probably own the rest. This leaves the founding owner with control, but gives the other key people a stake in the company's future and a right to be heard. It also provides security for the owners who don't have control.

In the early stages of its existence, a construction business might be organized as in Figure 9-7. At this stage, there may be only four people active in

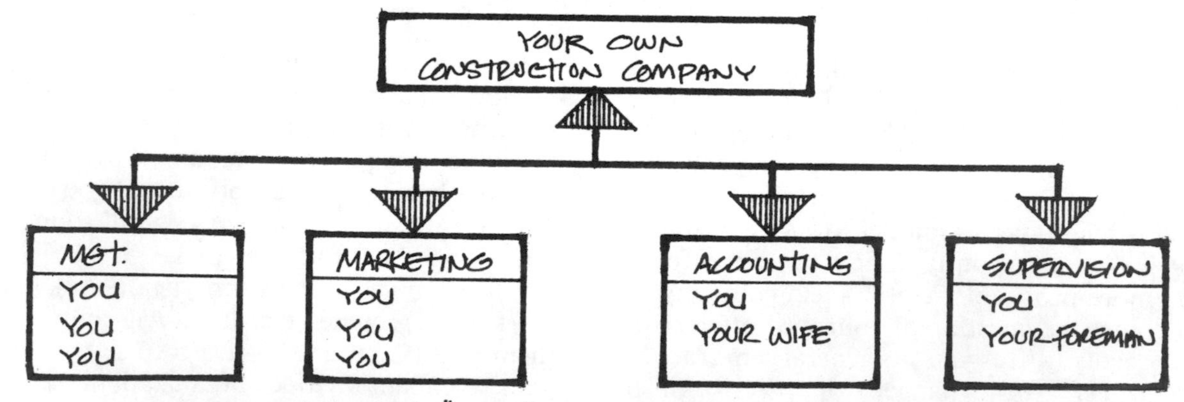

Small construction company
Figure 9-7

the company. Each has a distinct function and each may have an ownership interest in the company. As the company grows, organization will become more complex.

Your Family and Your Business

Organization of a family-run business will be different. Family-owned and operated businesses tend to be run like a family. And many construction companies are family-owned — maybe a majority. It isn't necessarily a disadvantage. Relatives tend to be loyal, long-term employees. Each knows what to expect from the others.

The danger in a family-run company is that business decisions are made for non-business reasons. The worst case is when some family member feels that he or she has a permanent job or an automatic share of income even if that person makes no contribution to income. That saps the energy of family members and discourages non-family employees in any business.

Every non-family employee has the right to resent the owner's nepotism. Everyone wants to compete on terms that he feels are fair and impartial. But having family on the payroll is always easier if those members were among the originators of the firm. As founders, they are more likely to be perceived as having earned the right to their position.

Summary

Define and set annual and lifetime goals. Begin *now* to prepare for retirement. Take a few minutes now to prepare a personal Finance, Goal and Time Projection chart like Figure 9-4. Develop a clear idea of the type of company you're building and where you're going in business. Pick out your specialty and stick to it. Remember, every construction company is like a chair with four legs: the originator, the marketer, the construction manager, and the accounting manager. Weakness in any leg can affect the whole company.

Going Where the Money Is

It takes three things to thrive in construction: generating profits, avoiding errors, and getting out of the way in declining markets.

Every builder has to learn to move assets and shift emphasis in harmony with the construction cycle. A profitable builder understands his market and anticipates the construction cycle. He sees opportunities and changes when they change. He relies on a competent team of professionals — tradesmen, supervisors, architects, engineers, and realtors — people who know their business and have the skills needed to perform their jobs.

Don't think of yourself as just a home builder or office builder or school builder or whatever. That's unwise. All building markets dry up from time to time. Be versatile. Stay within your area of competence, but plan to be where the money is.

Most builders make money in good times. It's the bad times that separate top builders from the herd. To thrive in the next building boom, you have to find a bridge over the abyss of a declining market.

In the good times, don't be too greedy. Leave a little in reserve. Avoid deals where everyone will lose if the cycle turns too soon. Remember that cost overruns get worse, not better, as you go through a job. There's usually no way out of a bad deal once it's been made. Avoid making them in the first place. When work gets scarce or construction is heading for the tank, listen to your calculator and your professional team, not to your ego.

What Goes Up May Not Come Down

Every contractor has to deal with inflation. Think of inflation as erosion of the value of money. By itself, it's neither good nor bad. It just makes things different. It's like a new deal of the cards in a poker game. Some players come out better and some come out worse. If you're on the wrong side of inflation, you lose as surely as if someone cut a hole in your wallet. But catch the inflation wave just right and your assets increase in value much faster than the dollar erodes.

Inflation tends to accelerate the pace of construction, because borrowers and buyers of hard assets are winners as the dollar loses value. That makes inflation a temptation for builders. Some get greedy. They want to buy or build everything in sight. But inflationary acceleration never lasts forever. Eventually deceleration crushes those who couldn't get out of the way in time.

Don't put all your eggs in one basket, even during strong inflationary periods. Avoid massive speculation, even when increasing inflation is clearly on the horizon. Be satisfied with controlled growth. Expand with systematic buying and building. Then plan on a period of settling-in and

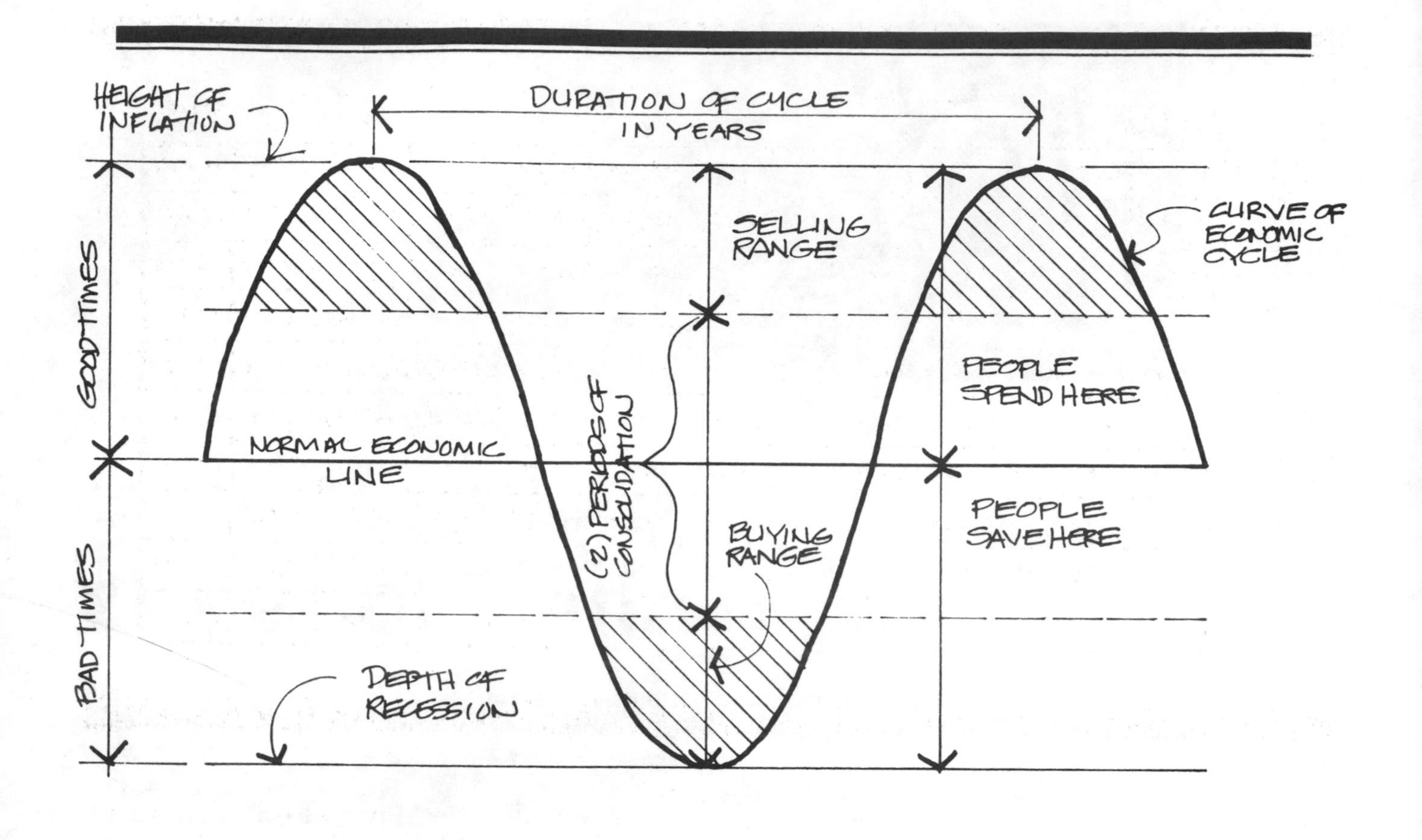

Buying, selling and consolidation
Figure 10-1

consolidation. Be comfortable with what you already own before expanding further. Keep some assets and borrowing power in reserve, even when inflation is accelerating. Never commit all your assets to a single make-or-break deal. Try to anticipate the market, not react to it. Recognize that the construction industry follows a three- to five-year cycle.

Look at Figure 10-1. Notice that we have one period of buying, one period of selling and two periods of consolidation. Try to have the properties you need to position yourself in the seller's market at the height of the inflationary curve. The best way to do this is to buy land and houses during the buying range at the depth of the recession. Of course, this takes cash and foresight. I'll talk a little later about how to amass the cash reserves you need. But you'll have to supply the foresight. It's not that hard if you pay close attention to the economic indicators. Read the business section of your newspaper and one or two good business magazines.

Any builder who ignores market cycles is going to have trouble. Building the same number of spec homes year after year is a prescription for disaster. Your market's mobile. Supplying a fixed product to that market leaves precious little margin for survival.

The Multiplier

Inflation is the multiplier. It's a terrific invention. It's so utterly simple that most people either don't understand it or overlook it. Here's the Rule of Multiplication: It's not what you own that counts so much as *how many of them you own.*

Look carefully at Figure 10-2. Notice that there

NUMBER OF HOUSES OWNED	ORIGINAL INVESTMENT	NUMBER OF OWNERSHIP YEARS AT 5% INFLATION 1	2	3	4	5	6	7	8	9	10	TOTAL PROFIT
1	50,000	52,500	55,125	57,881	60,775	63,814	67,005	70,355	73,872	77,566	81,444	31,444
2	100,000	105,000	110,250	115,763	121,551	127,628	134,010	140,710	147,746	155,133	162,890	62,289
3	150,000	157,500	165,375	173,644	182,326	191,442	201,014	211,065	221,618	232,699	244,334	94,334
4	200,000	210,000	220,500	231,525	243,101	255,256	268,019	281,420	295,491	310,266	325,779	125,779
5	250,000	262,500	275,625	289,406	303,877	319,070	335,024	351,775	369,364	387,832	407,224	157,224
6	300,000	315,000	330,750	347,288	364,652	382,884	402,029	422,130	443,237	465,398	488,668	188,668
7	350,000	367,500	385,875	405,169	425,427	446,699	469,033	492,485	517,109	542,965	570,113	220,113
8	400,000	420,000	441,000	463,050	486,203	510,513	536,038	562,840	590,982	620,531	651,558	251,558
9	450,000	472,500	496,125	520,731	546,978	574,327	603,043	633,195	664,855	698,098	333,033	283,003
10	500,000	525,000	551,250	578,813	607,753	638,141	670,048	703,550	738,728	775,664	814,447	314,447

The multiplier
Figure 10-2

are two multipliers at work here. One is the amount of time you hold a property. The second is the number of properties you hold at a given time. The most important multiplier is the quantity multiplier. Without it, the time element doesn't matter. You can have 20 houses for a year and make $2,000,000. But if you have no houses for five years, you'll come out with nothing.

Let's do an example using Figure 10-2. If you buy six houses, costing $50,000 each, and hold them for eight years, the total value at the end of the eight years would be $443,237. (See shaded box.) This is a profit of $143,237 ($443,237 minus the $300,000 you paid for the houses) over an eight-year period. But you only made a 20% down payment on each house, so your out-of-pocket expense for the six was just $60,000. If you take the profit of $143,237 and divide it by your actual expense, your gross eight-year profit is a whopping 239%. To find your annual return, just divide $143,237 by eight years, giving you an annual profit of $17,905. Now divide that by your original investment of $60,000. You made a 29.8% annual return. Not a bad deal.

Housing prices have doubled every five to seven years since World War II. Therefore, to become a millionaire, all you had to do was buy twenty average-priced homes and hold them for at least five years. Then sell. The profit would have been over one million dollars (current value) during nearly any period in the last 40 years. Of course, twenty parcels of real estate is a whole bunch. But owning and developing land has been among the best investments over the last 40 years.

You should have a piece of this action. Here's how: Hold onto 10% of what you build to get the full advantage of inflation. Take advantage of inflation by becoming a project owner, *but not necessarily a spec owner*. It's important to make this distinction. Simply stated, thriving begins with the multiplier, that is, ownership. Own some of what you build and ride the crest of the inflation wave. That's smart construction.

Many builders have made a good living in spec homes over the last 40 years. But generally, the buyers of those homes have done much better. They usually resold a few years later and made several times what the builder made. And with very

SHOWN IN THOUSANDS OF DOLLARS

INDIVIDUAL IN POSSESSION OF HOUSE	STARTING VALUE	BLDRS. 10% PROFIT	# OF OWNERSHIP YEARS AT 5% INFLATION										TOTAL GROWTH
			1	2	3	4	5	6	7	8	9	10	
BUILDER ONLY WILL SELL HOUSE *	$100,000	$10,000	—	—	—	—	—	—	—	—	—	—	$10,000
BUYER ONLY GOT FROM BUILDER	$100,000	—	5.0	5.3	5.5	5.8	6.1	6.4	6.7	7.0	7.4	7.7	$62,886
BUILDER AS THE OWNER ALSO *	$100,000	$10,000	5.0	5.3	5.5	5.8	6.1	6.4	6.7	7.0	7.4	7.7	$72,886

* DOES NOT INCLUDE TAX AND INTEREST WRITE-OFFS DEPRECIATION OR CASH FLOW FROM THE ACTUAL CONSTRUCTION WORK OR RENTAL INCOME ETC.

Buyer vs owner's profits
Figure 10-3

little risk, effort or investment. It's the property owner that has the advantage in the profits game.

Look at Figure 10-3. Notice that the buyer outdistances the builder by a wide margin in the total profits column. But the builder-as-owner does even better, with a grand total of $72,886. And this doesn't take into account taxes and interest write-offs, depreciation or cash flow from the construction work itself.

There's an additional advantage to ownership. Say you could make a 10% profit on the construction of a $100,000 commercial property or apartment building. That's $10,000. Normally you would have to pay about $2,500 tax on that profit. But suppose you take a $10,000 ownership interest in the property instead of taking the profit. As a co-owner, you deduct depreciation and interest charges from your taxable income, offsetting that $10,000 profit and reducing your tax proportionately.

If you hold a 10% interest in much of what you build, you're in the multiplication game. But avoid unnecessary debt like the plague. Just remember, building is great, but multipliers are better.

Their Buck *vs.* Your Buck

Some economists have suggested that access to credit is the foundation of most wealth. You probably can name several people who made their fortunes with "OPM" — Other People's Money. It's happened many times. To understand this principle, you need look no further than your own income and the source of most construction money.

First, let's look at your income. If you're like most builders, your personal income is between $25,000 and $75,000 a year. Your living expenses probably are between 50 and 110% of after-tax income.

If you have an annual income of $75,000 and living expenses of 50%, you can't possibly accumulate more than $37,500 a year for building and investment. Ten years of accumulations are needed to build a decent commercial building. Three years of accumulations would build a good-sized house. At that rate, you could build less than half a dozen buildings in your lifetime. Building for cash with no loan is prohibitively slow. The only alternative is borrowing. Thus we enter the realm of "OPM" by default. And that brings us to banks and bankers.

Banks and Bankers

The real function of a bank is to pile up money in convenient stacks so people can come in and borrow them. For this "piling-up," the bank charges a fee which also seems to pile up quickly.

Anything that piles up money and loans it out again is acting like a bank. The U.S. Government, Bank of America, insurance companies, pension funds, your investors, your employer, your partners, your friends, parents and your spouse's "cash stash" are all banks in a sense. No matter who the lender is or how much is borrowed, the principle is the same. Everyone in business uses other people's money. This is especially true of the construction business because construction is capital intensive. Banks spend the bulk of their time collecting money from depositors who come through the door of the bank. They pay depositors 5 to 10% and collect 10 to 20% on loans — all with little risk because of the collateral required for loans.

You can see that banks are middlemen. They earn a fee by piling up other people's money and lending these piles out again. Is there any way to bypass the middleman and save the middleman's fee? Could you stand at the front door of a bank and intercept depositors before they can get to a teller's window? Could you offer potential depositors higher returns and the safety of a bank if they invest with you? In practice, no. You can't offer most depositors all the bank services they need. But don't give up too soon. Some depositors may find your offer very appealing. And they may have a pile of money big enough to do you some good.

Believe it or not, it's possible for you to arrange your life and business affairs so you can work exclusively with investor money, thereby eliminating one of your major debt partners, the banking institution.

Some builders have learned to work without banks. Not only is it possible, it's desirable. Remember, banks don't like to be flexible. They want regular payments on a specific date — whether or not you collect your receivables or make a profit. A private investor is likely to be less interested in monthly payments. And private investors may be willing to adjust payments based on the performance of the project.

Though handling private investors has its own special problems, greater flexibility may more than offset any difficulties.

The Limited Partnership

For the builder who's reluctant to borrow or can't borrow as much as he needs, the limited partnership is a good alternative — either for part or all of the funds you need. Limited partnerships deserve careful consideration from every speculative builder. I'll cover the high points here and outline some of the ways you can get in trouble.

Understand first that limited partnerships have a bad reputation with some investors, and in some parts of the country. LP's have been used occasionally to swindle investors and dodge both state and federal taxes. But don't let that discourage you. It isn't the form of organization that's bad, it's the organizers. If you've got a good project that can succeed on its own, bringing it to the world in the form of a limited partnership won't do any harm.

But because limited partnerships have been abused so often in the past, all states (and the federal government) have passed, or are considering, legislation that would control LP formation or change how they are taxed. Be aware of the possibility of legislation, and know what the current laws in your state require before going too far with your plans.

Limited partnerships are like corporations in some ways. Limited partners are investors, just like stockholders. They have no day-to-day voice in running the company and are not liable for the partnership's debts beyond their investment. The general partner runs the store and will be liable for partnership debts. But limited partners have a big advantage over stockholders. Stockholders are taxed on dividends received, of course, but not on corporate profits. On the other hand, if the corporation loses money, none of the loss shows up on a stockholder's tax return (until the stock is sold or becomes worthless).

Not so for limited partners. The partnership pays no taxes. Instead, all gains *and losses* flow through to the limited partners. That's a big advantage if the partnership can be expected to show losses (from depreciation) in the early years of its life. Limited partners hope to have profits, eventually, of course. But when profits finally begin to show up, they may qualify for the lower tax rate that applies to capital gains.

Because the tax advantages of limited partnerships are so obvious, they have been both used and abused frequently. Today, all states set restrictions on how limited partnerships are formed. If your partnership has more than a certain number of investors, special disclosure requirements must be met. You may not solicit more than a handful of potential investors. The participants should be friends or close business associates, perhaps even family members. Your investors may be prohibited from buying in if they don't meet certain financial requirements. For example, net assets (exclusive of a home) must exceed a certain amount and yearly income may have to be above a certain amount.

Be sure to keep a current personal financial statement on file for each investor. State inspectors might one day ask to see them.

Of course, you can't advertise publicly for investors. Remember, you're supposed to know these people before they became investors. In many cases, this is only half true. Investment news travels like lightning. Many partnerships are composed of people who become fast friends and business associates only while the limited partnership is being formed. But no matter where you find your partners, the fact remains that you can't advertise for them. Doing so constitutes a public offering and is subject to the rules and regulations of the *SEC*, the Securities and Exchange Commission. You don't want to get mixed up with them.

Get Good Professional Advice

If you're considering a limited partnership to raise capital for real estate development, for heaven's sake get good professional legal and accounting advice. Nearly every metropolitan area has attorneys and accountants who specialize in the formation of limited partnerships. They'll be invaluable to you and your partnership's success.

The Prospectus

Limited partnerships are sold with a *prospectus* that outlines the purpose of the partnership, the money to be raised, and its use. The prospectus lists the general partners' names, addresses and their qualifications to handle something like the project being considered.

The purpose of a prospectus is to disclose everything potential investors should know. Keep some important information out of the prospectus and you've got a sure lawsuit if anything goes wrong or if the investors don't make as much money as they expected.

Distribution of the prospectus will usually be limited to about 25 copies. You can show it to only 25 potential investors. And only 10 investors will be permitted to participate.

Nearly every prospectus you see will be similar in some respects. The first few pages will have general information about the people involved, the property itself, buildings to be constructed, and so on. It will also include a cash flow and expense breakdown for the life of the project, usually 5 to 10 years. After that time, most of the heavy deductions will have been used up and the partnership will be sold.

The next part of the prospectus will outline the ownership percentages — who gets what and when: how profits and losses are to be divided. This is extremely important. Many times all the tax deductions are passed to the investors as added incentive to get into the deal, even though they don't own the whole partnership. Deductions, profits and losses are not necessarily tied rigidly, point for point, to the percentage of ownership. But make sure this is in your prospectus. If it isn't, the courts will invent a distribution plan, one that at least some of the partners won't appreciate.

Somewhere near the front of the prospectus there will be warnings about how the whole thing may collapse. You know that old saying, *Buyer Beware*. Every prospectus has plenty of that language.

If your project loses money and the investors get wiped out, that's O.K. — *if* the right warnings were in the prospectus. In practice, some promoters steal from their investors and get away with it because the prospectus explained that they were thieves — and disclosed that they planned to run off with everything that could be moved. That's legal if the investors gave consent on the dotted line.

The prospectus is designed to disclose to potential investors that their money is going into a risky deal that may go sour. If it does, they'll lose their investment. Language like that helps remove responsibility for losses from the backs of the organizers — and may help them win the lawsuits that usually follow when one of these deals goes belly up.

Your prospectus will include two more sections: the actual limited partnership agreement and a personal financial statement to be filled out by the potential investor.

The limited partnership agreement is almost always written by an attorney. The other parts of the prospectus you and your accountant can write. But the actual partnership agreement has all the legalese that only attorneys and judges understand. Leave it to your attorney.

The partnership agreement says who will keep records, file tax returns, where the business will be conducted, how general partners are removed, how sale of partnerships are to be handled, and how profits and losses will be divided. In short, it's the basic document that describes all rights and responsibilities of partners. Boring, but vitally necessary.

Here's another point to ponder: One or more of the organizers will usually be the general partner. The GP usually has an interest subordinate to that of the limited partners. In other words, general partners should participate in profits only after all limited partners have received distributions equal to their original investment, plus interest at a specified rate. Only after all of this has been paid out should the general partners take their share of the profits.

Limited partners consider this an added incentive that guarantees the general partner's best efforts. Limited partners also hold an interest in the land as collateral. Usually it can be sold to recoup at least 75% of their original investment. The land should be recorded in the name of the limited partnership as developer, not in the name of the contractor.

A few more words of caution before we leave the subject of limited partnerships. Generally, the offering may be made only in one state, and no commissions may be paid. But partners can move to other states after buying and fees may be paid to the seller for other services rendered. General partners should do the selling, not hired professional salesmen.

Running a limited partnership is simple. The general partners do everything, except put up the money. The limited partners do nothing *but* put up the money.

One note of caution: The general partners are responsible and liable for the well-being of the partnership and the investors' funds. However, just because a limited partner is described in the prospectus as a limited partner, it doesn't really mean that he or she isn't actually a general partner. The distinction between limited and general partner can change without the titles being rewritten. All that has to happen is for a limited or general partner to assume the role or duties of the other. If a limited partner begins to take on the activities of a general partner, then he or she may in fact become a general partner and have a general partner's responsibilities. A general partner may also become a limited partner by ceasing to act as a general partner and assuming the role of limited partner. But this probably won't relieve him or her of a general partner's liability or compensation.

The biggest danger in using limited partnerships is a stampede of unhappy, uninformed investors who want out and want out quick. A limited partnership isn't like a bank. Funds aren't available for withdrawal on short notice. Neither is a limited partner's interest like publicly-traded stock. There may be no market for a limited partner's share. That means the present value is nearly zero in a forced sale.

It takes a good, people-oriented diplomat to keep limited partners content. Not every contractor has the time, talent or interest to do that. Are you willing to sit through countless meetings and phone calls with your investors? It may take several formal and informal meetings a year to keep your partners content. It's a lot like herding chickens — a minute of inattention and they're off in different directions.

Your partnership will have to survive recessions, setbacks and misjudgments. You'll need cash reserves in the beginning to ride out the unexpected. Don't expect to go back to the limited partners for additional money after the partnership has been formed and recorded. It's happened, but don't count on your investors for a second round of financing. If more capital is needed, the general partner is usually forced to sell some of his own interest in the project.

Finally, remember that limited partnerships are business transactions, not your personal piggybank. Successful contractors who use limited partnerships protect their track record. They want a string of good deals with satisfied partners spreading the word. A few bad deals that end in lawsuits can put any builder out of the limited partnership business — permanently.

From any perspective, you have to work with the other guy's money. It's a way of life for almost every builder. But the way construction projects are bankrolled is changing. You should change with it.

Try to avoid using your own money on projects that involve partners or that are done for people other than yourself. Invest your money in your own projects, built for your own account exclusively. Don't get partners involved in these projects. This cuts your money risks to a minimum. Always isolate your money and risk wherever possible. Work on their buck, not your buck, when any risk is present.

Cash Is King

Thriving in the construction business requires cash. Builders who have cash will survive. Those who have more cash will thrive. Those who have a lot of cash can retire, take trips and count their money. But no matter how you cut it, *cash is king*.

There's no substitute for free cash reserves. That doesn't include borrowed money or the ability to borrow. It doesn't include the equity in your house, car, boat, equipment or business inventory. Neither does it include second trust deeds if you borrow against them — only if you sell them. Stocks and bonds are cash reserves if there's an immediate market for them. But when the market changes, so does their value.

The only real reserves are cash, gold, silver, or assets that are immediately salable, regardless of the economic circumstances at the time of liquidation. If you plan on thriving in the building market, start building real cash reserves. Believe me, you'll sleep better at night.

Without reserves, everything is a struggle. Buying on credit is always more expensive. Credit transactions increase the seller's cost and carry the risk of nonpayment. Every seller has to sell at a higher price when he sells on credit. The cash buyer should, and usually does, get the lowest price.

With cash on hand, you'll be able to take advantage of equipment and material discounts. That lowers your job cost and increases your profit margin with no additional effort. These savings are like "found money."

There's a point of diminishing returns, however, for cash reserves. As the cost of what you're buying increases, the percentage of your cash reserves needed to complete that purchase also increases. And there's a point where tax considerations become important.

Cash on hand should equal 10% of your total bonding capacity. For example, if your bonding capacity is $100,000, you should have $10,000 in reserve. This doesn't mean equity in real property. It can be in certificates of deposit, pension or insurance funds, or anything else that's reasonably liquid.

Of course, most builders can't keep 10% in reserve. Many builders never even come close. But maybe that's why about 50% fail in the first few years of business and many of the rest are unprofitable.

It's common for builders to struggle along with 5% or less in cash, snowing their bonding company by floating unrecorded personal loans that look like cash reserves. Few bond underwriters dig deep enough to uncover this kind of deception. Some would rather ignore it because bonding companies make money only when they write policies. So, after a cursory investigation, a signed affidavit of solvency and an updated financial statement, it's *voila* — one construction bond coming up.

Staying Power and Reserves

Your staying power describes how long you can hang in there as a builder when others are leaving the industry. Your staying power is determined by your cash reserves, liquidity and debt ratio.

Every business starts with a loss. Going into business costs money: a phone, office equipment, stationery, the first week's payroll, etc. This money has to come from savings, sold assets or borrowing.

The shaded area between the debt line and income line in Figure 10-4 is the money needed to carry the business during this start-up period. Hopefully, the debt line drops with time and the income line rises to meet it. The point where the lines intersect is the breakeven point. As income rises above debt, you bank the first real profits. That's the first opportunity to set aside cash reserves.

The magic minute in every business is when, through skillful management and debt control, assets actually begin to exceed debts. It's tough for most builders to get there. Some never make it. If it happens to you, don't blow it with new debt. Ignore that urge to buy the latest luxury car or a new tractor. Salt the cash away. Someday you'll need that money in reserve.

Unfortunately, too many contractors follow the performance curve in Figure 10-5. I call this the "Oblivion Curve." This builder started with the same business outlook as the builder in Figure 10-4. But poor management, inexperience, lack of sales ability and excessive debt sent him nose first into the Oblivion Curve. There's little hope of recovery short of bankruptcy.

A new construction contractor should be ready to supply from personal funds between 10 to 20% of operating expenses for the first 12 to 24 months in business. Another rule of thumb is that a construction business won't be profitable for the first 12 months but should be profitable after 24 months of operation.

Here's another guide for cash planning in a new contracting business. Debts of the new business

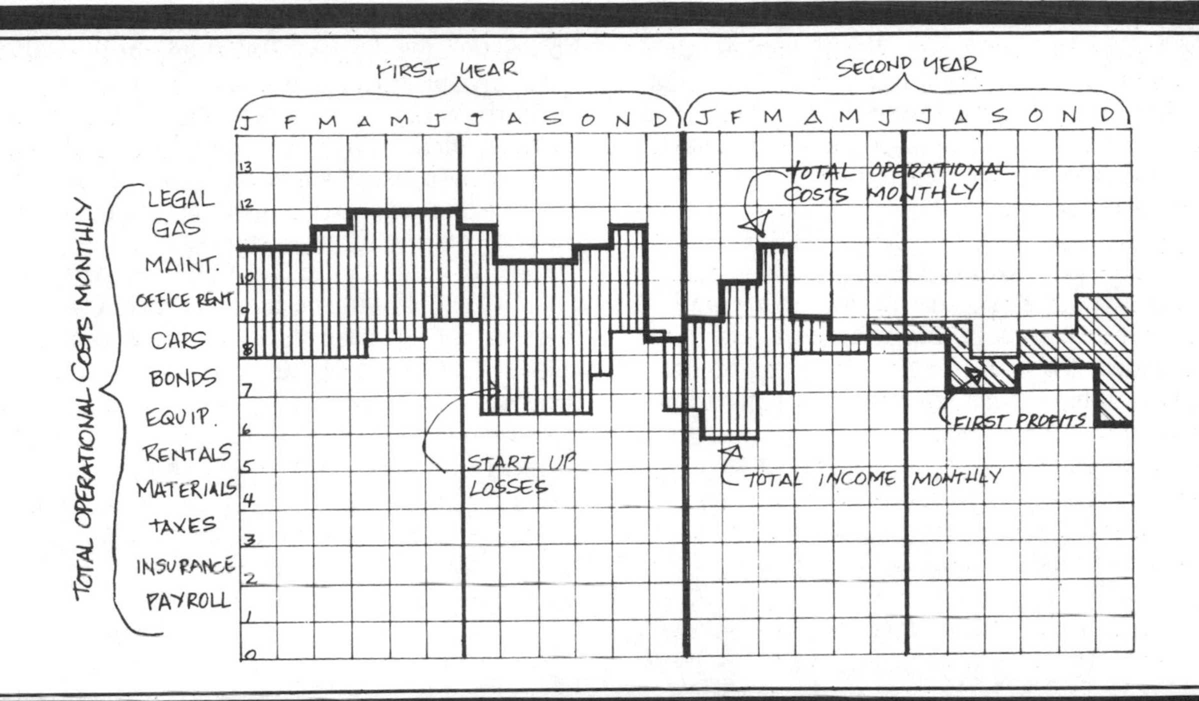

Start-up profits and losses
Figure 10-4

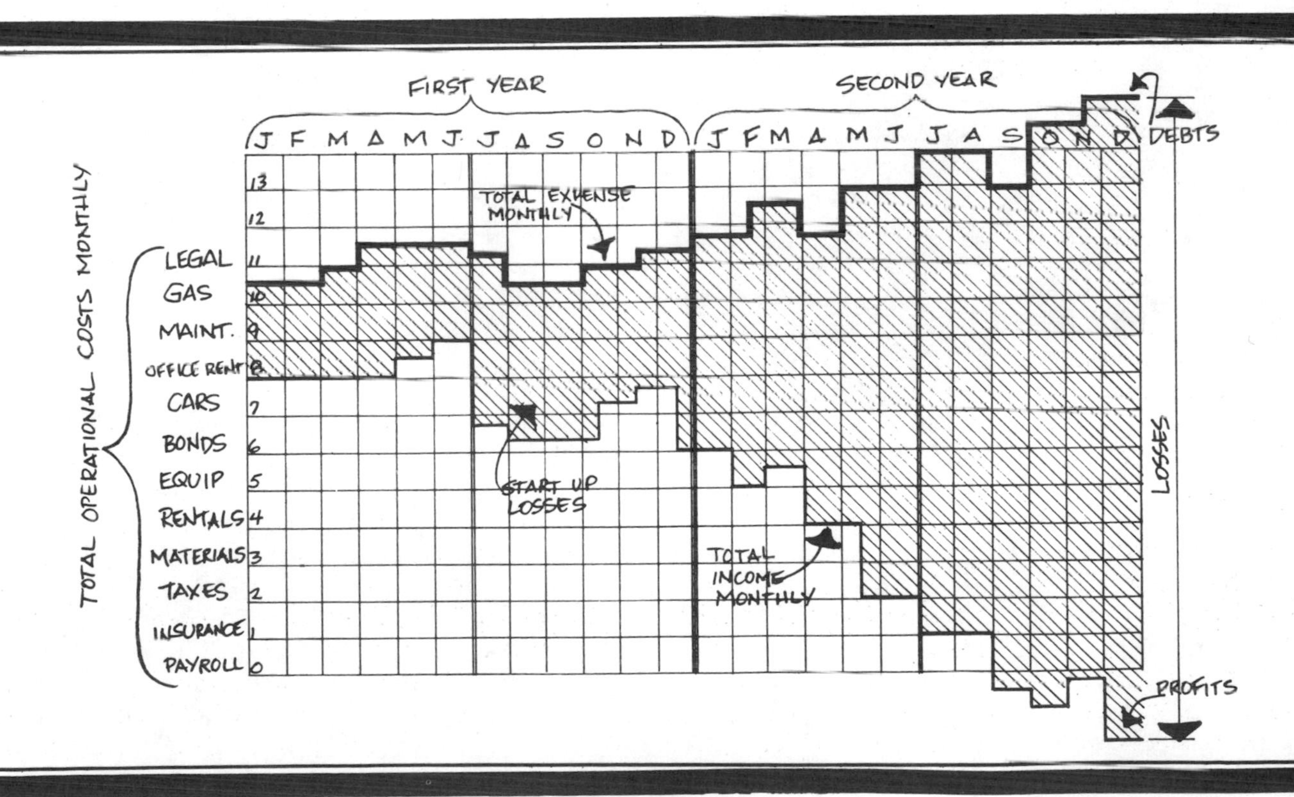

Oblivion curve
Figure 10-5

shouldn't exceed profits until the cash reserve is equal to 10% of bonding capacity. Even then, debt shouldn't exceed annual profits and 10% of bonding capacity combined. Eventually, debt load should decline. When it does, you're accumulating assets that can multiply with little effort.

Summary

Inflation can make you or break you. It's a matter of timing. Study the economic cycles in Figure 10-1 in this chapter. You can make inflation work for you. Remember, *quantity* is the important multiplier. It's not so much what you own — it's how many. Own 10% of what you build and let the economy inflate your assets.

Work with the other guy's money, not your own, especially when risk is significant. If possible, work with investor money and forget about banks. Try to generate enough cash to carry some projects on your own without partners.

Cash is king. Try to keep 10% of your bonding capacity available in cash at all times. Use anything in excess of 10% for material and equipment purchases. But don't dig into the 10% — that's your reserve.

So You Can't Find a Job?

The quickest way to shut yourself out of the construction business is to avoid talking to clients. If there's such a thing as professional suicide, ignoring clients is it. But still some contractors make that fatal mistake. They don't talk to their clients. What clients really want to know is what the costs are and when it will be finished. *Tell them!* Explain what you're doing and why. That's a very small burden, in my opinion.

It's the Sixth Time He's Called Today

It's a lot easier if you call the client first, before he gets the urge to call you. That puts you on the offensive, not the defensive. If problems are developing on a job, *bring the owner into it as soon as possible.* Keep him fully informed — on both the good and the bad. He's a successful man with money in the bank, borrowing capacity and good business judgment. Otherwise you wouldn't be building for him. He can handle any bad news you're likely to pass his way.

Your clients are *not* your enemy. I believe in that statement and you should too. Sure, there's an exception. I'll admit that a client becomes an adversary when he fails to pay a bill on time. But most billing disputes can be avoided if you provide back-up information that explains the bill in detail. If you've supplied that information and get paid, the client rates triple-A in my book. He has the right to be informed. That's both fair and sound professional practice.

If you think of yourself as a lousy salesman, maybe it's because you don't keep current clients informed. Poor communication earns distrust. And no one recommends someone he distrusts. Even worse, unhappy clients may be bad-mouthing your work to *potential* clients. Work that might have come your way without any recommendation may be going to others because of a negative remark by an old client.

There's no doubt about it, one unhappy, ignored client can unsell half a dozen future clients. If you're having a hard time finding and selling jobs, it could be because you've cut yourself out of the recommendation of satisfied clients.

Don't let your clients call six times before you call back. *You* be the one who calls *them* six times. You'll earn a good reputation and easy job leads. Every marketing plan begins with satisfied clients and a good reputation in the construction community. It's the easiest and cheapest self-promotion you'll ever do.

Public Relations and Your Company

There are two ways to handle self-promotion. The first is to hire a public relations firm. The second is to do it yourself.

Very few builders actually need a public relations

firm on retainer. Until you're the last builder in the country without a public relations firm, plan to create your own good relations. But you might want to interview several public relations firms anyhow. Ask them for fee quotes and a description of their services. From these interviews and descriptions, you should be able to glean enough information to start your own program.

A note of caution here. A poorly-focused public relations effort can do more harm than good. Don't present to the community a disjointed and incomplete picture of your company and your capabilities.

The Company Brochure

The function of self-promotion is to inform the community of your company and its business. I've found that one of the best ways is with a brochure that describes your company, its capability, and its achievements. We touched on brochures in an earlier chapter. Now I'll suggest what to include and explain the steps in developing a company brochure.

Don't even start making up your company brochure until you've done some research. Start with yourself. What type of business do you want? What kind of client are you after? It's only after you've made these decisions that you can make up a logical and well-focused brochure.

The Prospect List

The key to self-promotion is the list of prospects. That's your shopping list. It should include the name of the client, his dollar volume, contact person, phone number, address, number of projects built, type of construction (such as speculative or investment), the quality of construction and your assessment of how "hot" the prospect is. Flag the best prospects to make sure you don't overlook them. Keep a log of all contacts with each client, and follow up each contact with appropriate action. Figure 11-1 shows a prospect sheet with information about the potential client and a record of contacts. Set up something similar for your operation.

If you want to build for publicly-held companies, the information you need will be in the annual report put out by each company. Get these reports by writing to the company directly. If company stock is traded on a stock exchange, some brokerage houses will have annual reports available for the asking.

However you get a copy, the annual report will have a wealth of valuable information on income, size, market and products. Most important, you'll find a list of corporate officers by title.

You're looking for the Director of Facility Development (or some similar title), his address and his phone number. Compile a list of the corporate officers in your area that contract for construction services.

Next, use the civic and business organizations in your community. The Chamber of Commerce, real estate brokers, bankers and city managers know about building plans long before bids are solicited. These people can provide names and addresses of potential clients.

Put the address of each prospect on a card file and keep it current. These people will get your brochure. And they'll get it at regular intervals for many years. The list should include 500 to 1,000 prospective construction clients who can be expected to build in your service area in the future. That seems like a lot. But the list will grow quickly if it includes the name of every prospective client you meet or hear about. Add just two names a week and the list is over 100 at the end of the first year. Every time you bid on a job, add the owner to your list, even if you don't get the work. Anyone who builds this year is a good prospect to build in the future.

Designing the Brochure

Now you're ready to start designing the brochure. A word of caution. If you're a small contractor or just starting out, keep it simple and direct. The total cost of brochure artwork and printing shouldn't exceed two or three thousand dollars.

Get professional assistance with the layout and artwork. This is important stuff. Most typesetting shops can recommend freelancers who will design the brochure and lay it out for a few hundred dollars at the most. Better yet, find a contractor's self-promotion brochure you like and find out who designed it. Make yours similar or use the same artist.

Figure 11-2 shows the basic information that should be in your marketing brochure. For starters, you'll need the legal name of your company, your business address, phone number and license number. Next, list or describe the services you perform, your key people, and their background. Finally, list your references. Include a

MARKETING PROSPECT

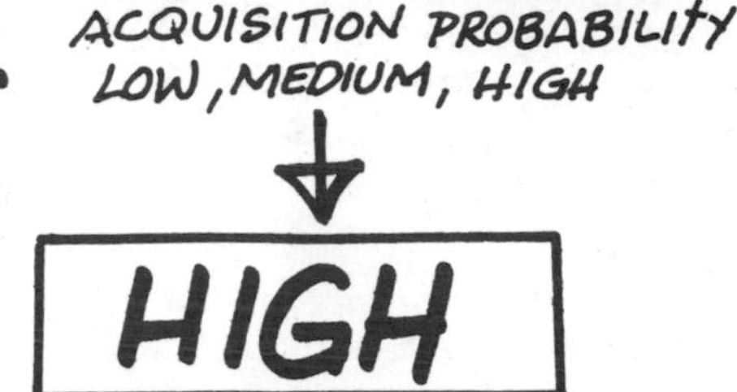

CLIENT: BIGTIME MFG. ENTERPRISES
1234 BIGTIME DRIVE
MONEY TOWN, USA

CONTACT PERSON: MR. BIG

PHONE: (714) 423-7000

TYPE OF WORK: FAST FOOD RESTAURANTS (INVESTMENT CONST.) WOOD FRAME, TILT-UP, AND MASONRY ON CONCRETE SLAB FLOORS, QUALITY MED. HIGH

ANNUAL DOLLAR VOLUME: $12-15,000,000 (25-30 FACILITIES)

CONTACT RECORD

DATE:	ACTION TAKEN	PERSON
3·18·86	MADE INITIAL PHONE CALL TO MR. BIG	BOB J.
3·22·86	MADE 2ND PHONE CALL TO MR. BIG	"
3·24·86	MR. BIG RETURNED CALL, TALKED, ASKED FOR INFO	"
3·25·86	SENT LETTER AND BROCHURE TO BIGTIME	BARB G.
3·30·86	CALLED MR. BIG, SET UP INTERVIEW FOR 4·17·86	BOB J.

Marketing prospect
Figure 11-1

couple of flattering photos of your work to make the piece attractive and give it flavor.

The photographs could be in black and white or color. But color adds at least $500 to the preparation cost and probably doubles the printing cost. Duo tones are a good compromise. A duo-tone print has black and one other color. It's made from a black and white photograph and adds little to the printing cost.

For format, I suggest that you stick to a single 8½" by 11" sheet printed on both sides but folded to 5½" by 8½". This makes a four-page document at a fraction of the cost of a printed and bound brochure.

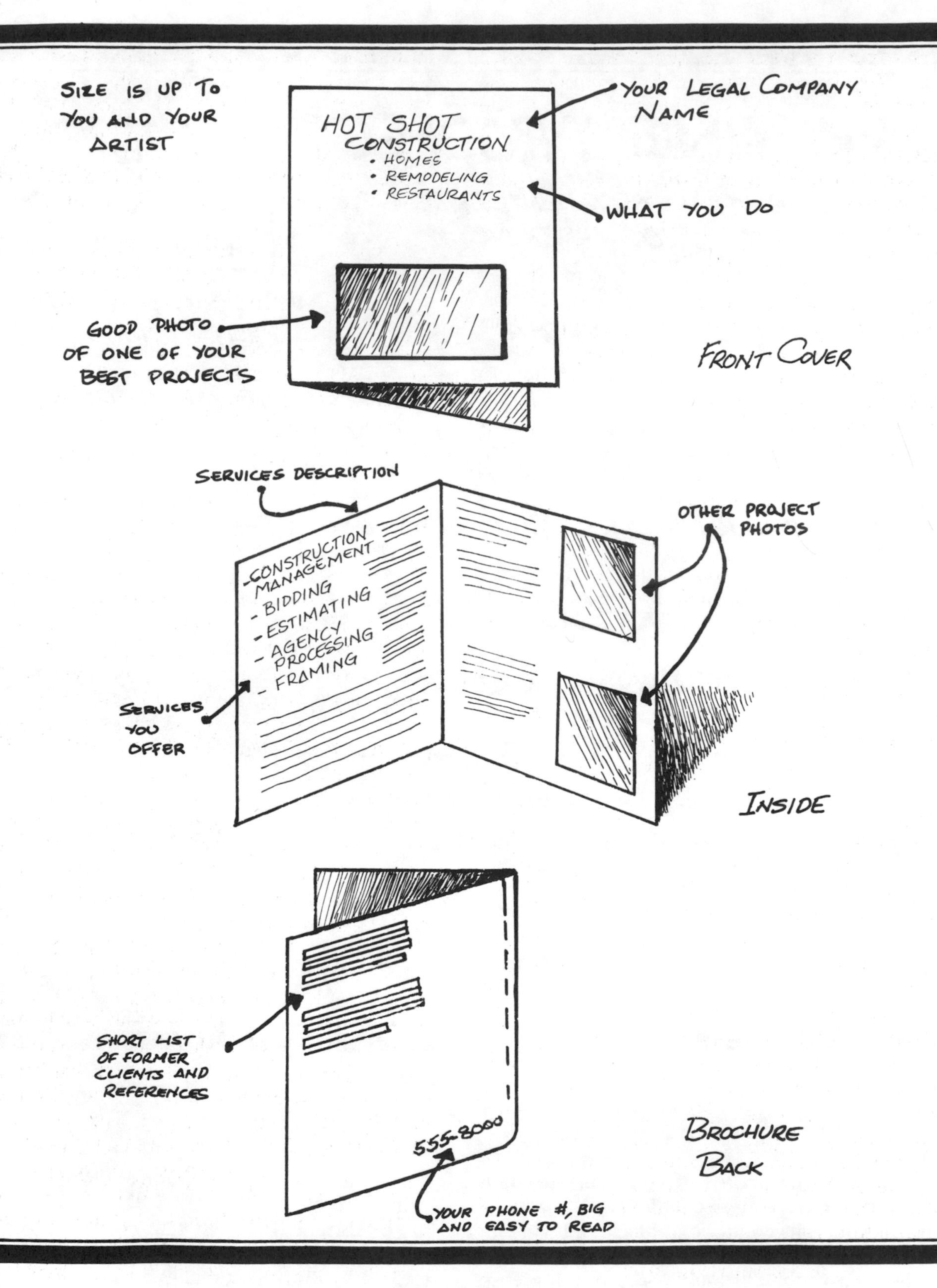

Construction brochure
Figure 11-2

The paper for your brochure should be card stock. "Ten point" is about the minimum. The cover of this book is twelve point card stock. If you use color, it should be coated stock, which has a slick feel. The cover of this book is coated stock. Also, consider colored or textured paper. It adds very little to the cost, but it may delay production a few days. Avoid brightly-colored paper. It's not professional.

Uncoated paper (like the pages of this book) have less glare and are easier on the eyes. But images look flat on uncoated paper. Printing inks tend to sink into the porous paper surface. If you'd like a slicker, more "uptown" presentation, use Chrome Coat paper. Chrome Coat comes in either a flat or glossy finish, but has a coating that confines ink to the surface, so images remain sharp and clear. Chrome Coat comes only in white. Other papers come in many colors.

If you decide not to use colored stock, use colored inks. Stick to complementary, muted colors for a professional look. But color isn't essential. I've seen many impressive brochures printed exclusively in black and white on glossy Chrome Coat paper.

Magazines and Newspapers

Local publications produce sales leads by keeping your company name in front of the public. Also consider trade journals or other types of specialized magazines. But keep track of the cost and record the number of actual leads and jobs this type of advertising produces.

My experience is that advertising in magazines and trade journals is an expensive way to stay busy. It's better to spend some time getting free publicity in a local paper or by distributing news releases. Introduce yourself to publishers and editors of local publications. What you build or propose to build is news. Local newspapers need to know about new construction in the community.

Recognize that all newspapers have an insatiable appetite for news. If the local paper has a real estate editor, he or she will be happy to receive stories about anything you consider newsworthy. Newspapers will even cover human interest stories that have nothing to do with your construction business. But be sure to include your firm name and describe your business.

Recommendations

I've always found the best source for leads in the construction business are architects, engineers, realtors and bankers. Start shopping for jobs here. But a referral from a professional like this carries an extra burden. If you make a mess out of the job, there are two reputations on the line, yours and the professional who recommended you. Nobody recommends someone who fouls up a referral job. Give it your best shot. Nothing counts like performance.

When you find work through the recommendation of a business associate, there's an unwritten obligation to keep that associate informed of job progress. Don't leave your associate in the dark. When the job's complete, take him or her to lunch. You need to say thanks. Then return the good deed by referring someone to him. Favors that business associates do for one another are the foundation of good business relations.

Flakes, Suede Shoes and Con Artists

Among your prospective clients will be a few individuals that are more likely to waste your time and money than produce income. I classify them as Flakes, Suede Shoe Operators, Con Artists and Poor Souls with Good Intentions. You should be able to identify these birds a mile away.

A Flake has about half the money needed to do what he's talking about. He assumes that you'll work for peanuts in the interest of the project he's proposing. He wants you carry the financial risk in return for his skill as deal-maker.

Flakes almost know what they're doing, but not quite. They have half-baked plans and want you on board to make it all work. But even your best effort won't make it work. And the Flake will lose interest and go on to his next victim long before anything concrete develops.

Mr. Suede Shoes is a different item altogether. These guys can usually be identified by their flashy clothes, gold chains and fancy cars. They carry a wad of credit cards and have no visible means of support. Mr. Suede Shoes doesn't know a joist from a girder, but as a self-appointed expert he's excited by the glamor of it all — by the big deals and the high finance.

Mr. Suede Shoes doesn't necessarily want you to put any cash into his project. Of course, he won't turn it down if you offer. But he wants you to donate all your overhead and profit. In addition, he'll ask that you run all over town helping him put his end of the deal together with the banks. He expects you to do his job and your job too. *Such a deal*. When all's said and done with the Suede Shoe developer, you'll find that more has been said than done.

Here's the best way to smoke out the Suede Shoe Operator: Ask about money. Insist on knowing the dates you'll get paid and the exact amount. And keep on asking until he either gives you a satisfactory answer or tip-toes quietly off into the sunset.

You won't run into the Con Artist very often, but he does exist. The construction Con Artist isn't like the Suede Shoe Operator. The Con Artist is a crook and he knows it. Mr. Suede Shoes just thinks he's a big wheel. The Con Artist is probably less flashy. He seems to be running a legitimate business and dresses conservatively. He wants you to put real money into a mythical project to be built on land he doesn't own. The projected profits will be big, *very* big. The problem is that the project either never existed or assumes loans that no bank would ever make. In our business, this is known as selling "blue sky."

The best protection, if you've met Mr. Suede Shoes or a Con Artist, is to run a credit check. Get a bank reference from him. Call the bank. They'll tell you in vague terms what the average account balance is for a specific customer. If he usually has $15,000 to $30,000 in the bank, they'll say that the customer keeps "a low five-figure balance." A "high five-figure balance" means that he usually has $70,000 to $90,000 in the bank. If the bank says the account is "unsatisfactory," that's all you need to know.

Any reputable business person should be willing to give you a bank reference. Anyone who won't probably has something to hide. Don't deal with them. Insisting on a bank reference will smoke out or scare off most Con Artists. You're getting too close to the truth.

And after investing in any project, run a preliminary title search to see if you've been recorded as the owner. If not, call the police and your attorney, in that order. Don't call the Con Artist. He'll skip out, leaving you and a raft of other investors with nothing.

Dealing with Mr. Good Intentions

You'll run into many people who think they can get a project built but lack the skill, the financing or tenacity to get it done. Many of these people use land development as a sideline to supplement their income. These poor souls have the will but lack the horsepower to erect the structure they have in mind. Even though they never get anything built, they spend thousands on detailed studies, estimates and schedules and can waste countless hours of your time in the process.

Don't be too critical of the prospective client who's sincerely interested in biting off more than he can chew. At some time we're all guilty of having eyes bigger than our stomachs. I know I am. I've let my ambition get ahead of my ability and my checkbook several times in my career. Many contractors take on more work than they can handle. When that happens, some jobs have to sit while others get the attention. Owners of the neglected projects are disappointed if not irate. Who can blame them?

Be fair with Mr. Good Intentions, but protect yourself at the same time.

My advice is to go ahead and spend a little time with Mr. Good Intentions. Sell him your services on an hourly basis. It's great for your cash flow. But if you're spending time with him and he's not paying, go on to somebody else who has a real deal. You'll never thrive in construction by wasting your time.

Credit Checks

One focus of your marketing effort should be to sort out the good clients from the bad. That's part of the marketing function — not just finding jobs, but finding sound, profitable jobs. There's no better protection than really knowing the people and firms you do business with. This is your best defense against being sucked into bad jobs by fast-talking developers and assorted sharpies.

For a small fee, a credit reporting agency such as TRW or Dun and Bradstreet will give you some important basic facts about a potential client. For instance, you can find out about foreclosures, bad checks, unpaid debts, bank affiliations and property ownership.

The length of time a potential client has done business in an area gives you an idea of his stability. And the longer a client has been active in a community, the more history there will be on him. If your client's new in town, check him out on his home turf. If the information you receive is sketchy, run a preliminary title search on the property he claims to own. Most title companies will do a preliminary search for less than $100. It could be the best $100 you ever spent.

Personally, I find making credit checks both tiring and distasteful. But if you're going to protect yourself, it's the best way.

The best credit information will come from people who've already done business with the individual in question. Ask other contractors, subcontractors and suppliers about a potential client. Ask about a client's payment record and track record in real estate development.

If you can't establish a client's sincerity, ask for a reasonably large retainer. If he balks, it could be because he doesn't have the cash. Suggest that his bank issue a letter of credit in your favor. If that's impossible for some real or imagined reason, back away from the deal.

Where's the Money Coming From?

I can't emphasize this too strongly. Understand the financial arrangements on every job *before* you start work. As a contractor, you should see where the money's coming from as clearly your client, if not better.

It doesn't matter how good a project looks or how much you'll get out of it if your client can't finance it. If he hasn't got the money, you haven't got the job, even if there's no competition.

Always ask these questions: "Where's the money coming from?" "Who's going to pay me and when?" Don't start ordering materials until there's a satisfactory answer. Your financial future depends on it.

If a lending institution will be the primary source of funds, notify them that you're involved in the project. Develop a personal relationship with the lending officer who'll disburse funds to you. Find out how the payment procedure works. That way you supply cost verifications and material and labor releases on time and in the form required.

The dates payments are due should be made clear right from the start. Set up a payment schedule at the beginning of each job, just as you do a job progress and completion schedule. The payment schedule is based on the construction progress schedule. Once you know how much work is to be done each month, you can predict when payments will be required.

Work out the job completion schedule and payment schedule with the owner and then submit it to the lender for ratification. The schedule may be either simple or complex, depending on the scale of the project and the degree of detail required. Obviously, the more detail, the more complete and accurate the final results will be.

The payment schedule in Figure 11-3 is a simple chart showing the amount of income expected from a particular job for each month of the coming year. You'll want to back it up with a written payment schedule indicating the major phases of the project, the time each phase is to be completed, and the amount due for the work done in each phase.

Inclement weather will always extend the construction period and may increase costs as well. Your construction schedule should include an allowance for weather days. How much time to allow will depend on your location, the kind of weather you have, and the time of year or calendar duration of your project. No matter what estimate you make for weather delays, give yourself some elbow room, especially if there's a liquidated damages clause in the contract.

Another type of delay to consider is delays in payment. A draw that isn't available when expected can cost you dearly in penalties and lost discounts. Here's an example. On a $20,000 lumber bill, loss of the 2% discount for prompt payment will cost you $400. That's not much compared to the total construction cost. But if the job went for $80,000 and your profit is 5%, the loss of that discount is equal to 10% of your profit. That's nothing to sneeze at.

Evaluate the chance that you won't get paid on time and allow for the higher cost of late payment.

Your estimate should also include an allowance for *contingency*. Construction is seldom done for less than expected. Anything unanticipated usually raises costs rather than reducing them. The contingency allowance covers things that can't be foreseen. Of course, it shouldn't be used as a substitute for careful estimating. But putting an extra 2 to 8% in a bid for contingency is considered good practice by most estimators. On a small commercial or residential job, an allowance of 6 to 8% would be reasonable. If you work with any less, you're either an exceptional estimator or fairly foolish.

Summary

Good relations with your customers are the foundation of your business. You don't need a professional public relations firm. But be your own booster. Someone has to promote your services if you want to succeed. Use marketing brochures and free news coverage in local publications to keep your name in the public eye. The newer your com-

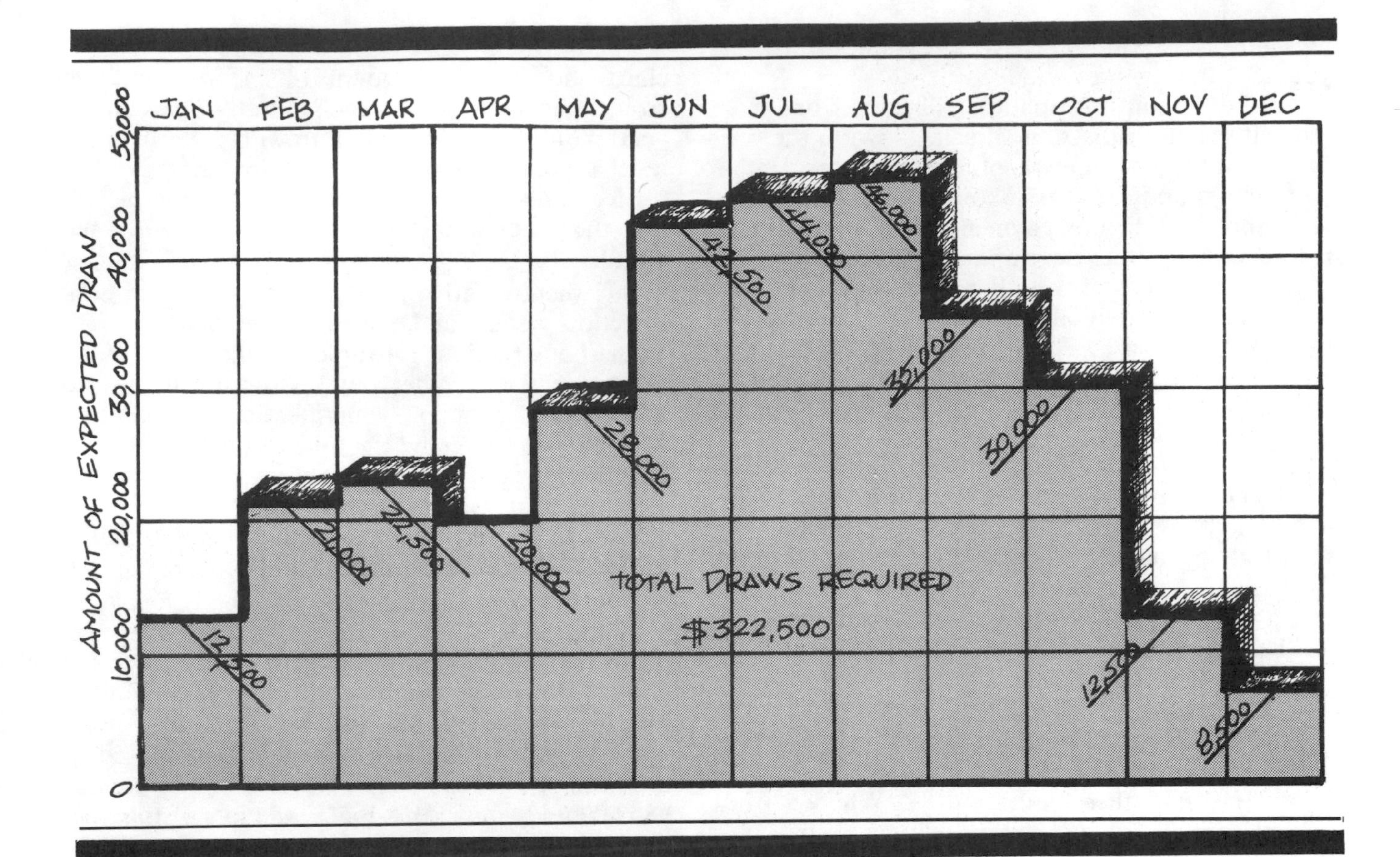

Time payment schedule
Figure 11-3

pany, the more important this is. What potential clients and the public learn about you is your responsibility.

Flakes, Suede Shoe Operators and Con Artists — the woods are crawling with them. The trick is to keep them at arm's length. That's not as simple as it sounds. The easiest way I know of to separate the wheat from the chaff is to talk money and payment dates right up front. If they don't have the cash, they're wasting your time.

Good intentions only count in fairy tales. In business there are bills to be paid, choices to be made, and limits to be set. Good intentions alone are worthless. It takes money, skill and perseverance to complete projects on time and within budget.

Over-Design, Under-Design & No Design

There's more to building than just construction. Dealing with design professionals (architects and engineers), and government, is an important part of construction contracting and land development. Most successful builders spend a lot of time working with design professionals and government agencies. In fact, the more successful you are, the more time you'll have to spend in architectural and engineering offices and with government bodies. If you feel more comfortable driving nails and bidding jobs than negotiating with architects and commissioners, this chapter may be valuable.

Architects and Engineers

Everything you build should be built from professional-quality plans. Fortunately, most architectural and engineering firms prepare adequate construction plans and specifications. On some jobs, the plans may have gone through the plan approval process and have been finalized before you even see a set of prints. You, as the builder, may have little influence on design of the final product. Your only responsibility may be to execute the design as conceived by the owner, architect and engineer. On other projects, you may help select the design staff and even participate in the design process.

No matter what your role, most builders have discovered that architects and engineers don't think like builders. They don't see many of the construction problems in the plans they draw. Because architects neither pay for nor build what they design, they don't have to be practical. They're in the aesthetics business, not the construction business.

Engineers tend to focus on what will do the job and what won't. An engineer's primary goal is finding a design adequate for the intended purpose, getting you from where you are to where you want to be as directly as possible. The architect wants to get you there, but his goal is getting you there in the best possible manner. Quality of the journey is what's important — not how quickly it can be completed. Architects create the concept for a project; the engineer tries to make it practical and buildable.

Engineers are nuts and bolts people. They work with strengths, loads and capacities: drainage, grading, streets, walls, floors, roofs, dams, water systems, sewage disposal, bridges, power distribution and air-conditioning systems. The engineer's seal on the plans is his certification that the design meets generally-accepted load standards.

In the order of creation, the owner begins the process with perceived needs. The architect proposes a form of building that will meet those needs. The engineer finds a way to make the architect's concept meet load and strength standards. Finally, you, the builder, must find a way to turn that design into a building. This is the traditional pecking order in the construction industry.

On some projects you may have the luxury of selecting your own architect. Some construction companies either have an architect on staff or have established working relations with one or two architectural offices. If you're selecting an architect for the first time, interview several in your area that handle the type of projects you build. Over lunch, do a little brain picking. Find out what they've done and how they approach their projects. Ask how their office is organized. What's their specialty? Who's on the staff? Find out how they get work and what they charge for their services. Within an hour, you'll know if their operation fits with yours.

A key qualification for any architectural or engineering firm will be their ability to work with you. Remember, the street should run two ways. You'll send business their way and they should be willing to refer business to you.

You may want to select more than one architectural firm — a small firm for small jobs, a larger firm for larger jobs.

Don't be afraid to inquire about fees. That's an important part of your investigation. There are recommended fee schedules for both architects and engineers, but the fee is always negotiable. Smaller architectural and engineering firms are less likely to follow recommended fees. A large firm on a complex project may charge more than the recommended fee.

Fee schedules are based on the estimated cost of construction. Notice this. The higher the cost, the more the architect makes. Is it any wonder that architectural firms include construction details that drive builders up the wall?

Never assume that an architectural or engineering firm has your financial interest at heart. Like you, they may be deeply involved in your building's structure, design and construction. But their perspective is different. Even an architect on your payroll needs some supervision to keep the cost and schedule from getting completely out of hand.

When you can participate in the design process, watch the work carefully. Review the plans and details as they are committed to paper. Most excessive costs in a project arise during the planning and design stages, not in construction. The best and easiest place to begin cost control is when your project's still on the drawing board.

Zoning and Permits

Getting your zoning changes and permits can be a real pain for the inexperienced. The best advice I've heard is to find which way the grain is running and then try to run with it. Conform with the city and county master plans whenever possible. Nearly all cities and counties have already prepared a master zoning plan. Get a copy and review it with your client, architect, engineer and attorney before even considering a request for variance.

Most master plans have many points in common. Generally housing is clustered in one part of town and industry in another. Between these two section falls everything else: stores, office buildings and apartments.

Look at Figure 12-1. Notice that office and apartment buildings are used as buffers between shopping centers or industrial and residential areas. Notice the use of waterfront areas for parks and public recreation purposes.

Don't try to build a manufacturing plant in a residential zone. The plan was prepared by school-trained professionals who take their work very seriously. The commission that adopted the plan isn't going to look with favor on any significant change in what it perceives to be a good plan. They'll spot gross incompatibility like a skunk at an ice cream social.

Here's a capsule summary of my advice on rezoning and zone variances. Don't even bother unless there's a good argument that the change is in the city's best interest. If not, forget it.

The whole idea in construction contracting is to make more money with less effort and risk, not the other way around. Take my advice. Get to know your city planners and administrators. Study the city's master plan. Select property locations that are compatible with the city's objectives, as well as your own. Don't waste time and money fighting uphill battles over ill-chosen property.

If there are good reasons for a change in zoning or for special consideration in granting a permit, get professional help to prepare your case. Line up some heavyweights on your side. But be selective. Not all architects, engineers and attorneys are experienced in dealing with public agencies.

If you decide to request a variance or special consideration on a permit, approach the department staff with a healthy degree of respect and patience. The last thing you need is an enemy at City Hall. Don't let some ardent out-of-town owner push you into anything that would damage your

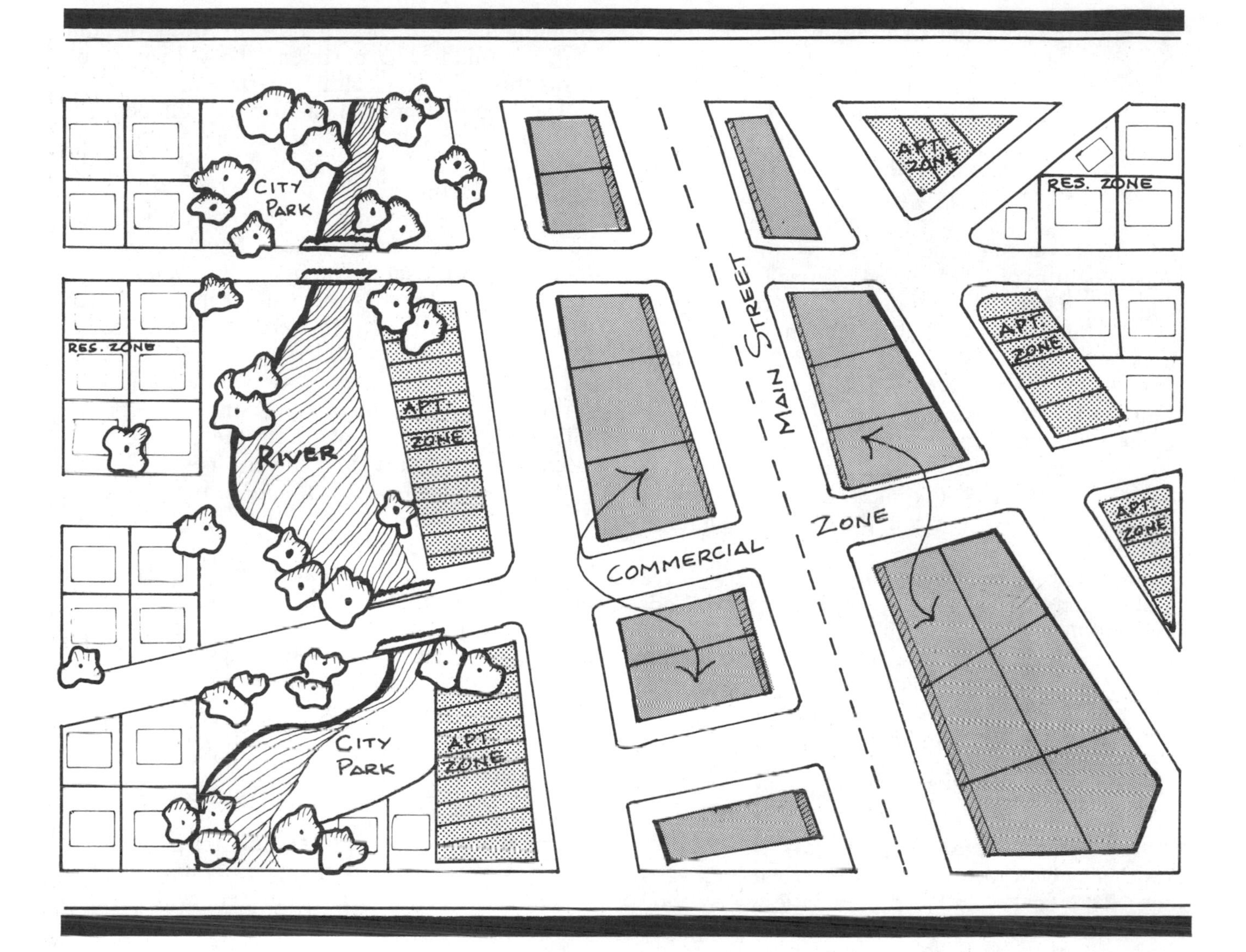

City master plan
Figure 12-1

reputation as a responsible member of the community. You have to work with those same city officials year-in and year-out. Your client doesn't. Believe me, they have an elephant's memory when it comes to troublemakers.

Here's the approach I use. At the outset, ask the city or county officials for the low-down on all approvals, permits, testing and certifications necessary to complete the project. Then start working with architects and engineers. Don't assume that they already know what you know. They probably don't. Even if they do, what they know is at least partly out of date.

If necessary, ask that the city's staff meet with the owners, architects and engineers. Clarify the process to everyone. Let the city know what you're planning and get their advice on the best way to get approval and the time required. Try to discover the snakes in the woodpile before they bite you. Don't make any significant investment until you know the rules that apply.

If you have a meeting, take careful notes. Type up the notes and distribute a copy to everyone involved. It's amazing how effective this will be in helping everyone recall what was discussed and decided. Otherwise, you may find that what you think was agreed on is different from what the others think they agreed to.

Planning and Building Departments

Every builder should distinguish between the responsibilities of the Planning Department and the responsibilities of the Building Department. Try not to get the two confused. You have to take the right problem to the right place at the right time in the construction process.

If you take questions about parking requirements and public access to the Building Department, you've exposed yourself as a rank amateur. Those are Planning Department problems. But questions about occupancy permits or life safety systems won't be answered at the Planning Department. That's the business of the Building Department.

The Planning Department passes judgement on issues about property usage such as setbacks, height restrictions, building function, materials, traffic control, loading, and the environmental impact of a project.

The Building Department is concerned with the public safety. They review construction methods and building systems, including mechanical, electrical, and fire sprinklers. They also pass judgement on anything relating to structural requirements, heating, cooling, roofing and exhaust systems, ramps and handicapped facilities.

If that's clear, let's move on to the other city and county agencies you'll have to deal with.

Other City and County Departments

Another agency you'll meet is the Health Department. They review and inspect facilities where food is processed or prepared for public consumption. They also have authority over medical and drug-processing facilities, all water wells, and disposal facilities.

Typical concerns of the Health Department include sanitation conditions, air conditioning, range hoods, grease traps, sinks, drains, ceiling, wall, and floor finishes, paint colors, tile textures, water temperatures, equipment types, closures, bases and accessibility.

My advice is to bring the Health Department into the picture early if your project requires Health Department approval. Schedule a meeting with them. You should be there, of course. Also include your client, the architect and the interior designer. At the meeting, get a copy of the filing requirements that apply to your project.

In some cities and counties you may have to file plans with both the Health Department *and* the Building Department. In others, you'll file only with the Building Department. They'll route your plans through the Health Department.

The Health Department in many smaller cities won't have the staff or experience needed to review plans. Actual checking may be done by the County Health Department.

Another stop on your tour through City Hall will be the Fire Marshal's office. The Fire Marshal focuses on two issues. The first is the availability of water to fight fires. Is the water volume, pressure and location adequate if there is a fire? Second, they decide if fire sprinklers and smoke detectors are needed. If so, they pass on the location of sprinkler outlets, control valves and shutoffs. The Fire Marshal's office also sets the distances between fire hydrants and determines where the hydrants will be located. Be sure to review the Fire Marshal's decision. Putting the hydrant in a bad place will cost you plenty.

The Public Works Department is an interesting little agency with some rather large responsibilities. Public Works oversees streets and street widths, parking lots, sewer lines, water lines, storm drains, bridges, dams and other public systems. If your project involves anything in this area, you'll need to visit the Public Works Department. Your civil engineer is the professional most involved with this agency. If you need help deciphering their requirements, ask him.

The sewer, water, power and telephone companies have their own engineering staffs and requirements. You'll deal separately with each of them.

Finally, there's the agency that provides public transportation. The Transportation Agency sets standards for street turnouts, turnarounds and even street widths in some cases. Also, they can veto power line locations. So call them before going too far with your work.

In short, there's a maze of agencies and departments that builders have to deal with. Learn your way around early in the game. You can't make a profit while you're groping your way through the maze, trying to find the exit.

Plans and Specifications

Some builders never have the opportunity to participate in the design process. That's too bad. My experience is that input from a construction contractor during the design phase can cut costs and reduce the construction period considerably.

If an owner-client invites you to review some plans before they're finalized, or if the designer is working for you, make the most of it. Be ready to provide the insights and suggestions that only an experienced builder can provide.

Start by keeping the plans and specs simple, well-organized and to the point. Make them idiot-proof. The field people who use plans aren't always college graduates. And they're certainly not mind-readers. Make your instructions as clear as humanly possible. The fundamental idea behind all plans and specs is communication. Don't load your plans with obscure details and boilerplate language that doesn't really matter. Make them models of clarity.

Plans and specs serve two distinct functions. The plans (working drawings, construction prints, or blueprints) describe how it all goes together — how the various products and materials are to be assembled. Specifications describe the materials to be used, the conditions under which the work will be done, and any tests the materials have to pass.

Here's where you, an experienced builder, can help the designer. You know what type of work goes up quickly with less waste and little trouble. You know which methods work best, cost least and take less time. Suggest that the designer avoid details that will be more difficult to complete or that involve some risk from your standpoint. Recommend materials that are easier to work with or are more readily available. Favor construction that can be done with your own crews and with no special tools. Discourage work that requires subcontractors and special equipment. Following your suggestions should cut costs by 5 to 15%, while shortening the project duration.

Then go one step further. Talk to the architect and engineer about the content and assembly of the plans and specs. Review the list of drawings and specs to be prepared. Eliminating redundancies, conflicts and unnecessary effort should save 5 to 10% of the design cost.

Your help with the specifications should save the architect lots of time. You already know what materials are appropriate for the type of building being designed. That saves design time. The architect should pass these savings on to the project owner. Keep in mind, though, that architects and engineers like to make money too. They're not going to show you how to cut their fees. It's up to you to make the suggestion and show that the suggestion is valid.

Here's another trick to reduce the design cost. Set a reasonably short time limit for preparing the plans and specs. Have the designer agree to a deadline before giving him the job. The shorter the period of time available to prepare your documents, the more efficient your designer has to be to complete the job on time. Of course, setting a tight deadline shouldn't force the architect into overtime work. That defeats the purpose. Apply just enough pressure to assure continuous and efficient effort without overtime work.

Pay special attention to all the major pieces of equipment your designer selects. Design people usually don't know the cost of equipment until the bids are opened. The designer's choice is seldom based on a cost comparison. Propose equipment that fits into the construction budget, meets code requirements, is compatible with other systems and has adequate capacity.

Never allow the owner and architect to canter through the design process unbridled without an appreciation of costs. Failure to focus on costs in the design stage is sure to bring in bids that are over budget. That wastes everyone's time with redesign and more estimates. It probably delays construction several months. Meanwhile, inflation boosts costs a couple of percent.

Finally, recognize that both over-design and under-design are hazards in this business. Although over-design is far more common, some architects don't do enough drawing to completely describe the work to be done. Other architects insist on making minor or even fairly major changes through the working drawings, bidding and even construction. An architect who over-designs is adding costs that aren't necessary. An architect who under-designs or can't make up his mind can cost the contractor all of his profit, and more.

Summary

Before you choose your design professionals, spend some time getting acquainted. It's time well spent. When you have a choice, work with architects and engineers who are sensitive to a contractor's viewpoint.

Zoning and permits are headaches for all builders. Try to conform to the master plan. Avoid going against the grain. Get to know the city planners and administrators. Let them know that you're a responsible, quality-conscious builder. Know the rules before you start building and keep everyone on your construction team informed. Don't hesitate to schedule meetings with officials in the departments involved. City and county officials are paid to have meetings — all sorts of meetings. Give them the low-down on what you're doing. Find out what they need from you and do your best to comply.

Work closely with the designer whenever possible. Begin your cost-control effort while the project's still on the drawing board. Be ready with suggestions for cutting construction costs without reducing quality or utility. Suggest alternatives in materials, equipment and scheduling — whatever will bring down the overall cost of the project. No one is better at cost control than an informed and alert builder.

Second, for the Third Time

What Business Are You In?

Work on a fee basis whenever possible. Make it the lifeblood of your construction business. If all legal work went to the lawyer with the lowest bid and all construction was done on a fixed fee, builders would drive Mercedes and lawyers would drive pickup trucks.

But if the market you serve runs on competitive bidding, it's survival of the fittest. You'd better take a close look at the competition. What kind of firms are you up against? Are they big or small? Are they specialists or general contractors? How many are there in your area, in your specific line of work?

If you pick up a bidders' list on a project and find 35 listed, you're in trouble. Unless you know something that the others missed, your chance of success is less than 3%. Odds like that don't make anyone rich. In fact, you probably don't have much chance of breaking even on that job.

To live on competitive bids, you'll have to improve your odds significantly. If the odds are less than one in four, forget it. You'd be better off playing golf. The simplest and most direct method of improving your odds is to develop special skills or knowledge that let you compete in an area that isn't already overcrowded.

Here's an example of what I mean. I have a contractor friend who does quite well building small concrete dams, equipment bases, water channels and pumping stations for public agencies. Here's why he's successful. There are only three or four contractors in his area that bid on work like that.

My friend used to handle residential foundations, concrete slabs, sidewalks and retaining walls. He had to compete with every joker in town who owned a pickup truck and trowel and thought he could pour and finish mud. With 800 slab finishers running around loose, he found himself donating his time on most projects.

My friend had to make a change. And he did. He's still in the concrete business, but his clients are well-financed municipal governments. Sure, the paperwork is more demanding and the job specs more complex. But government tends to pay bills on time and in full. His competition has been reduced to a handful of firms. That makes his pricing policy more realistic. My friend makes a good living and feels good about his business and his future. That's important.

Here's the moral to my story. If you're going live on competitive bids, find yourself a solid, profitable niche. Otherwise, your best bid will come in second 90% of the time.

Of course, it's not easy to find that niche. But there are probably over 500,000 small construction markets and specialties in the U.S. — including many that still wait to be discovered.

Where do I get the half million figure? Start with the three broad areas of building construction: residential, commercial and industrial construction. Then add government work, remodeling, insurance repairs, and non-building construction (drainage, power lines, etc). That's seven major construction categories. You can probably think of others I haven't mentioned. Then multiply those seven categories by the 15 or 20 subcontractors that service each of these major categories. That makes more than 100 major construction specialties. Then multiply by the 4,000 to 5,000 communities in the U.S. that are large enough to support a specialized construction contracting business. That gives us between 420,000 and 700,000 options available when selecting where and what you're going to do as a construction contractor.

Of course, many of these markets are saturated already. There's no need for another builder with the same skills, equipment and knowledge. But other markets are under-served. And there's probably a reason why. Either the work is too remote, requires too much sales effort, too much equipment, too much paperwork, too much capital, too many trades, or has some other disadvantage that's discouraged others so far. But some bright young contractor is going to discover the market and find a way to overcome the disadvantage that's discouraged the others. That contractor has stumbled onto a gravy train.

And once in a while a brand new market opens up. It could be because of sudden growth in a community, such as a new factory locating in the area. Maybe a new construction material has come on the market (like elastomeric roofing), requiring special skills and equipment for installation. Perhaps there's been a change in the type of buildings needed (think of the growth of the fast-food restaurant industry, for example.) Contractors who guess right about the new market and are ready to serve it first will become established and profitable competitors as the market expands.

No matter what the reason, don't keep slugging it out with the competition in an over-served market. Look for an under-served market or growing market that gives all competitors a chance to thrive. And be ready to overcome whatever disadvantages come with that business.

If Bid You Must . . .

Here's the first rule of bidding, if you haven't learned it already. *Always bid what you've been asked to bid.* Don't elaborate or qualify your bid without specific written direction from the architect, owner or contracting official.

The quickest way to get yourself cut out of a bid opening is by not bidding what you're asked to bid. The owner wants to compare apples to apples, not apples to oranges. If the plans and specs call for painting the exterior walls with catsup and a toothbrush, then that's the way you should bid it. Maybe you feel cherry-flavored bubblegum applied with a nailfile is the right way to go. You have a perfect right to suggest that procedure. But bid catsup with a toothbrush, even if you press the argument with whoever has authority to change the specs.

The bid package will identify who can change and interpret the plans and specs. All suggestions and questions should go to that person. Usually this will be the architect. But the owner can fill this role, as can an engineer, a construction foreman, a construction manager or a secretary. It doesn't matter who's handling the job. What does matter is that you address the right person and give him time to respond.

It's usually to your advantage to respond to the "or equal" clauses in the specs and suggest where savings are possible. If your suggestions are adopted, you gain a distinct advantage over the competition. Here's an example: The specs call for 1,000 chrome-plated hookplates. You have a source for 1,000 vinyl-coated hookplates at $10 apiece cheaper than the chrome model your competition is pricing. If you can get a change approved, there's a $10,000 bidding advantage. And that's the name of the game — the bidding advantage. Figure out a better, quicker, cheaper way to build and you pocket most of the savings.

Every time I pick up a bid package, I begin by looking for anything that's to my advantage. Search through the specifications for the "or equal" clauses. Point each one out to your subcontractors. Look at the alternates on the job. Many contractors have submitted the lowest base bid only to end up $6.03 higher overall when the alternates are considered.

Look over the architects' and engineers' detailing. There's usually a way to simplify most fabrication and erection. When you find opportunities for savings, point them out to the owner if you must. But keep what you've found under you hat. That's your bidding advantage.

Bid Peddling— Bid peddling is endemic in the construction industry. Owners do it. Contractors do it. I don't think it's either fair or good business. But sometimes what I think doesn't count much.

Here's how it works. An owner gets a bid from you. He then offers the job all over town to anyone who will do the work for a dollar less.

Before criticizing that practice, let me remind you that general contractors are notorious for bid shopping. After a contract has been awarded, it's common for the winning general contractor to ignore the subcontract bids that won him the job. Instead, he solicits new subcontract bids or even offers the work to other subs at a negotiated price. Many general contractors feel that this is their right. I won't argue. But it doesn't breed loyalty among subcontractors. And no general contractor can get very far without competitive, reliable subs.

Open bidding helps curb a client's bid-peddling instincts. But nothing will stamp out the practice completely. Short of threatening a lawsuit, nothing stops an owner from approaching the next highest bidder — or another, uninvited bidder — and suggesting a price that would give him the job. The only practical way to protect yourself from bid-peddling is to know who you're bidding for.

Of course, you're nearly always protected from bid shopping on government work. The bidding procedure is established by law. That advantage encourages many contractors to handle *only* government work — city, county, state and federally sponsored projects. In fact, the largest firms do government work almost exclusively. Of course, clean bidding isn't the only reason to specialize in government work. Governments have the right to tax. That creates deep pockets and perpetual financing. Government agencies burden you with requirements that seem insane. But they won't peddle your bid and they will pay up.

Extras— Every extra a contractor claims probably seems like price gouging to most owners. But construction is too complex to think of everything before work begins. And it's too permanent to ignore the changes someone wants to make during construction. So there will always be extras — and there will always be disputes over what charges are legitimate.

Any time you have to add something that isn't on the plans and specs, submit a bill to the owner. These are justifiable charges. The owner has to pay. That's why it's important to watch all extra costs. Bill out your extras at the end of each month. Include documentation such as a copy of the owner's request, time tickets and material receipts.

Include in the charge all extra costs, not just the cost of labor and materials. Be sure supervision, overhead, taxes, insurance and a reasonable profit are included in the extra charge. It's foolish to handle extra work at your cost or below your cost. Extras have to carry their share of all overhead.

Here's how important it is to collect for extras. Say you have a $100,000 job and did some work not shown on the plans. Your cost for the extra work was $1,000 including overhead but no profit. You expected a $10,000 profit on the job. You're debating now whether to hit the owner for the extra $1,000. If you don't, your profit on the job is down to $9,000. That's a 10% drop. In effect, failing to charge for the extra is like doing $10,000 worth of work at no profit. And the slimmer your profit margin, the more important every extra becomes.

Collecting for extras gets tougher and tougher as time passes. That's especially true when the job goes over budget, even if you aren't to blame. The only way to get paid for all extras is to document every charge to the teeth. Good paperwork wins claims for extra charges — especially if you have to go to court.

Protect your profits. Don't let enthusiasm for the job or preserving good client relations prevent you from collecting for extras. Every extra not charged comes right out of your profit. There's really only one shock absorber in any business, and that's profits. Nothing else is as adjustable. Everything else is a fixed cost which must be paid. Profits alone suffer the penalty for mistakes and miscalculations. When the profits are gone, the reason for doing the job is lost. That's the way the system works.

One final note about extras. Don't inflate your billing. Be firm but fair. Gouging an owner is a good way to lose repeat business. At best, you'll lose a recommendation. More likely, you'll end up in court.

Negotiated Fees and Design-Build Contracts

Many smaller projects are now built on a negotiated-fee basis. It's an informal procedure that helps both builder and owner. Both can adjust prices and the work to meet economic and budget conditions.

Here's how it's supposed to work. The owner

selects three or four general contractors on the basis of their merit and reputation. He has each evaluate the plans and prepare an estimate. The estimate is for information purposes only, although it may include firm subcontract bids. The contractor then sets a fee for the project that covers only his supervision, overhead and profit. Construction is done at the cost of labor and materials plus the negotiated fee. The most enterprising contractor with the lowest bid on his fee will probably get the job. The unsuccessful contractors are compensated for their time and effort. This arrangement is fair to all parties involved. Of course, negotiated fees aren't a panacea for bad construction management. If you were a lousy estimator before you had a negotiated contract, you'll still be one when you land your first negotiated job. Neither is the negotiated-fee arrangement free of drawbacks. For example, owners can negotiate with several contractors at the same time. This presents a perfect opportunity for bid peddling.

If you enter into a negotiated-fee competition, I'd suggest that you prepare yourself for a financial free-for-all. You'll be selling not only price, but your skill, experience, staff and financial capability. A major consideration may be the time you can carry expenses before the owner has to pay the fee. The depth of your pocket can be an important issue.

Competitive bidding really demands less from a general contractor. He picks up his plans, assembles his price, and submits it to the owner. When the bid's opened, he either wins or loses. Nothing counts except the price.

Generally, negotiated-fee contracts will be profitable. There's less competition and a smaller chance that you're just spinning your wheels. In a negotiated-fee competition, you should get one out of three or four jobs. If you're invited into a negotiated-fee competition, give it a try. You just might find it's your cup of tea.

The Estimator's Art

Estimating is a learned art. Construction estimators are made, not born. Don't be discouraged if estimating seems like hard work. Everyone learns from the bottom up, through trial and error. And don't think you're a failure if you blow an estimate every now and then. Every estimator has. Mistakes are inevitable. It's how quickly you learn from mistakes that counts.

Estimating is an A-B-C operation. Start with Task A. It's probably the first thing you do on the job. Break Task A down into its parts and find every cost in every part: labor, materials, equipment, waste, supervision, delivery, tax, insurance, overhead, and fees. Then go on to Task B. When all tasks have been estimated, add up the totals and tack on a reasonable profit. That's pretty simple. But it isn't easy. It takes time — time to collect all the prices, time to be sure nothing is left out, time to check and recheck the figures.

What isn't simple is determining the labor cost — the time needed to install each item multiplied by the cost per installation hour. Estimating installation times takes judgement, knowledge of construction procedures, and cost records on previous jobs. If you can estimate labor within 10%, if you take the time to track down material costs, and if you use checklists to avoid overlooking any important item, the rest of the estimating is routine.

Good estimating is an essential part of construction contracting for most builders. So let's assemble a construction estimate. Look at Figure 13-1.

We'll use steel for our example. Find the steel sizes on the plans. Look through the specs for the type of steel needed. Call a steel yard. Give them the sizes and type. They'll quote a delivered price, tax included.

Next, you'll need to figure the beam, pipe, column and cap. Be sure that flanges and column caps are pre-drilled for the necessary bolts. There are 16 holes to be drilled. Also, column caps have to be welded to the pipe column. All of this drilling and welding should be done by the steel supplier or a local fabrication shop. Be sure it's either included in the quote or covered by a separate quote.

You'll also need 8 bolts as described in the detail. The material cost is easy. Just call a supplier. But you have to guess the time required for assembly, and compute the hourly cost for labor.

Finally, add the cost of the hoist, temporary supports, and shims. If you've covered all costs including shop priming paint, your estimate for this portion of the job is complete. The total cost is nothing more than the sum of the parts. But the only accurate way to find that total is to estimate each part separately.

But there is a good way to speed up your estimates on repetitive items. The slow way is to figure labor for 15 assemblies and material cost for 15 assemblies and other costs for 15 assemblies and then combine all costs. The faster way is to figure all labor, material and other costs for one assembly and then multiply by 15.

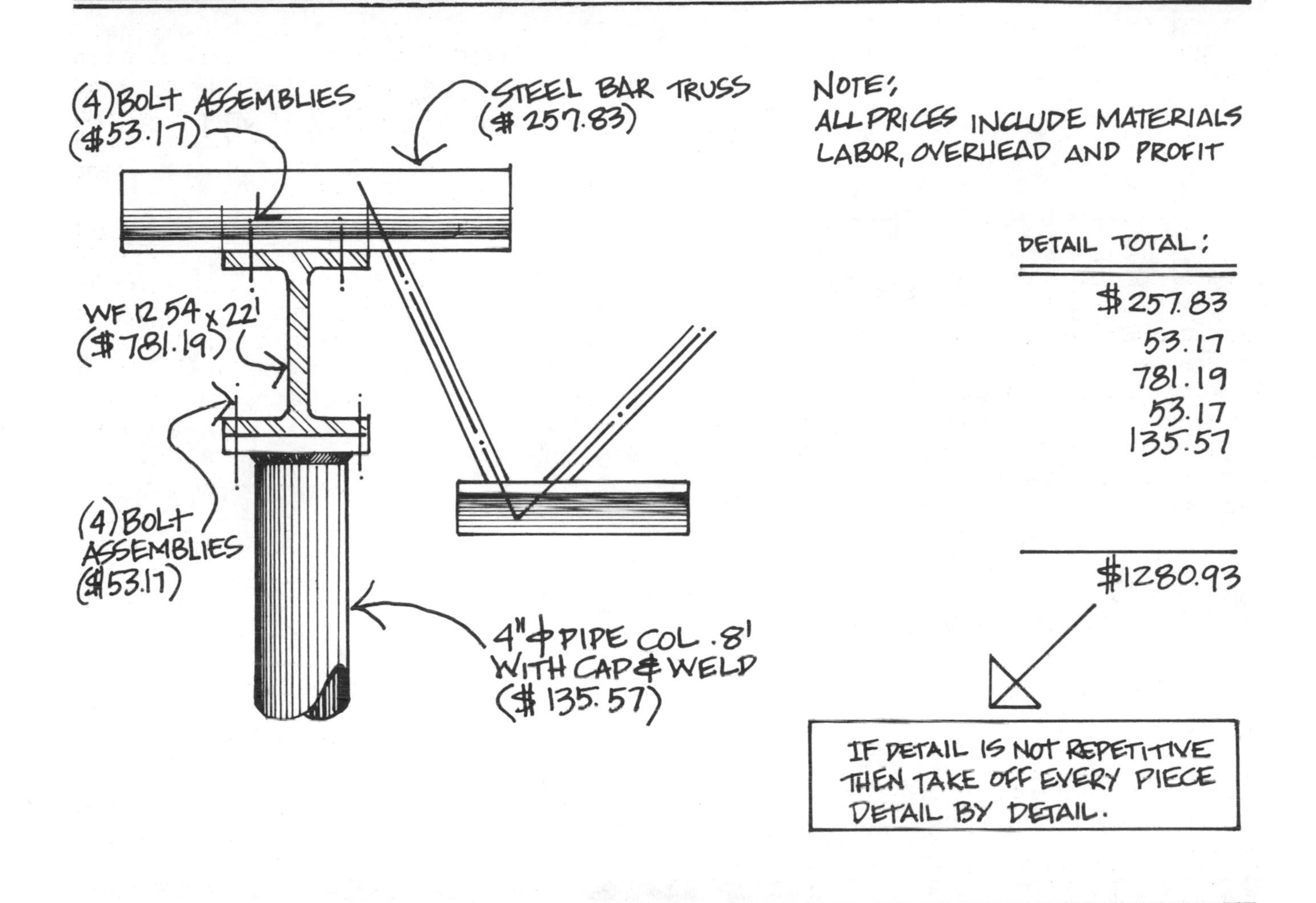

Basic estimating
Figure 13-1

Figure 13-2 is an example of the fast way. Figure the cost of one connection. Count the connections that are identical. Then multiply the quantity by the unit cost. That gives you the total cost for this item. Repeat the process for each detail shown in the plans. This procedure works equally well for lumber, steel, concrete, glass, roofing, electrical, plumbing, earthwork or planting. Obviously, the units change — from cubic feet to board feet to square feet — depending on what you're estimating. But the procedure is the same. Break each piece down to its parts, figure the cost of each part, add up the parts, and multiply by the units needed to get the total.

Adding Taxes and Insurance
Be sure the "labor burden" is included in your wage costs. If you're paying a carpenter $10.00 an hour (including all benefits), your out-of-pocket cost per carpenter hour will be about $12.50. The extra $2.50 is the "labor burden." It covers taxes which all employers are required by law to pay, and insurance which all employers are required by law to carry. There's no legal way to avoid this extra cost, and every estimator is well advised to calculate it carefully and include it in every bid. Here's where the money goes:

Unemployment Insurance— All states have unemployment insurance programs. The federal

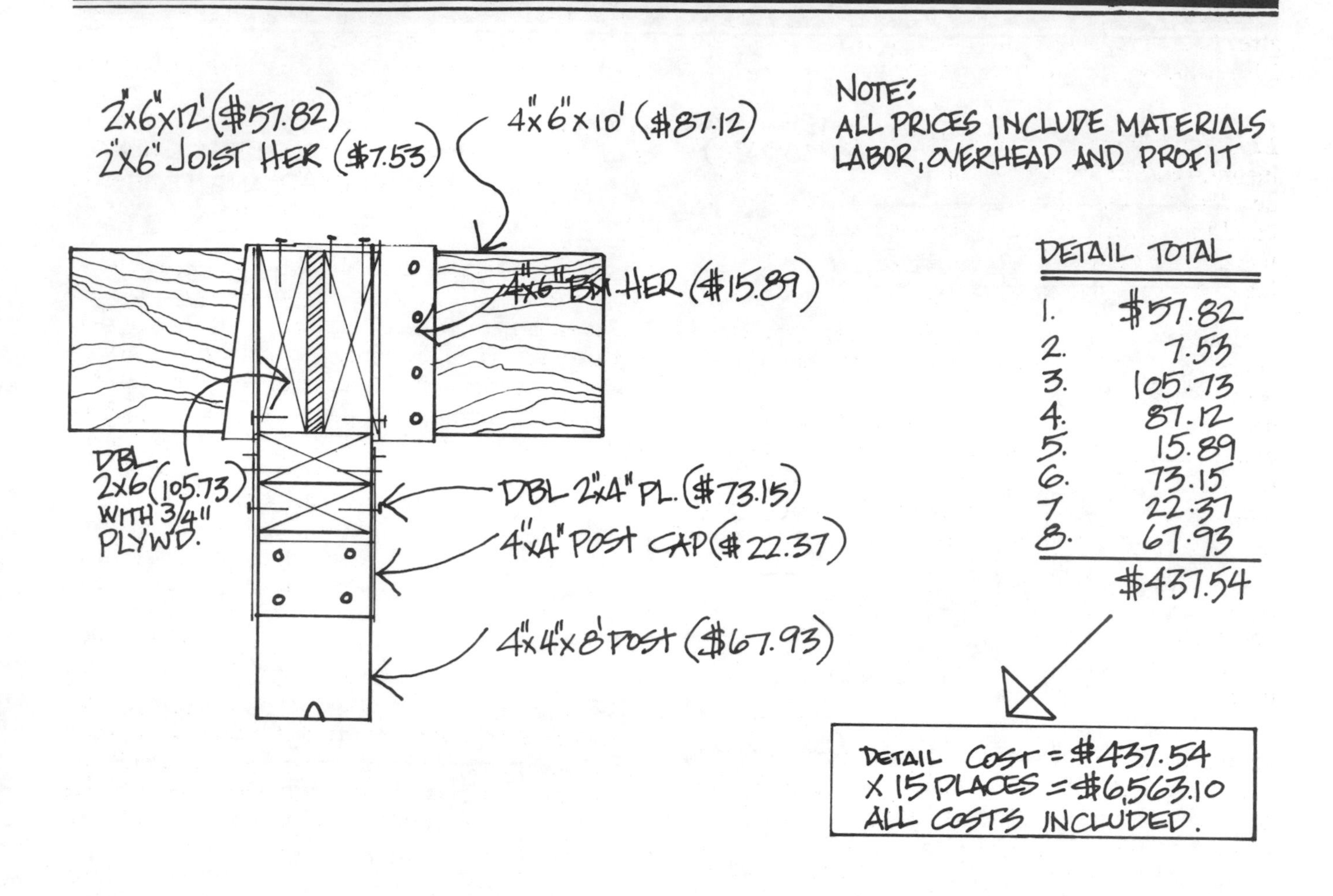

Fast estimating (by whole details)
Figure 13-2

government also has an unemployment insurance law. You know it as FUTA. Rates vary from state to state and from year to year. But your state probably requires payments equal to about 4% of payroll. The federal government's share is about 0.6%. For each dollar of payroll, you as an employer have to kick in an extra 4½ cents. So far, not too bad! But let's keep going.

Social Security and Medicare— Employers have always had to make contributions for Social Security and Medicare. You know this as FICA. For every dollar paid to employees, the employer has to pay about 15 cents to the IRS. But the employer's share is only 7½ cents. The other 7½ cents is withheld from employee paychecks. That brings our labor burden up to 12 cents per dollar of payroll.

Workers' Compensation— Every state requires that employers provide workers' comp coverage for their employees. If an employee is hurt on the job, workers' compensation insurance covers the bills. The cost is only pennies per hour for office workers. But it's big bucks for some construction trades. Rates vary by trade, state and year, but the cost on most trades is between 5 to 15 cents for each dollar paid in wages. Your insurance carrier will quote the exact current rate for each of your tradesmen.

Adding the workers' comp cost to the taxes already covered, our labor burden is now close to 20%. And we're still not through.

Liability Insurance— Every builder needs liability insurance to protect his business in the event of an accident. Again, the cost is based on payroll. A good liability insurance policy with adequate limits will cost you about 2% of payroll. That brings our labor burden close to 22%. That hurts!

Understand this very clearly. Except for their share of FICA, employees don't pay these taxes. The money has to come out of *your* checking account quarterly. And there's no legal way to avoid paying this labor burden in full. States and the federal government impose heavy burdens on employers who don't make their deposits on time or fail to provide compensation insurance.

Don't fail to include the labor burden in every estimate. The best way is to add an appropriate percentage to each wage to cover taxes and insurance. Then use that labor cost when figuring all installation costs.

Overhead in Several Flavors

The last lines of every estimate are reserved for overhead and profit. It's my feeling that overhead and profit should be added as lump sum items at the end of the estimate. But I've seen them added into the hourly labor rates so every labor cost in the job includes both overhead and profit. Do it either way you like, but keep it simple.

Note especially that no estimate is complete without a line for overhead expense. Your profit margin shouldn't cover overhead. Overhead is a separate item entirely. And there are two types of overhead on every job: direct overhead and indirect overhead.

Direct overhead is sometimes called job overhead. It includes costs that aren't associated with any trade or material installed. Instead, they're the direct result of taking on a particular job: barricades, bonds (bid, completion, maintenance, street repair), building permits and fees, business licenses, clean-up, design fees, equipment rental, estimating fees, insurance (liability and casualty), job shack, job signs, protection during construction, repairing damage, sales taxes, sewer connection fees, street closing fees, supervision, surveying, temporary fencing, temporary utilities (phone, water, electricity), travel expense, watchmen, and water meter fees.

Estimating direct overhead is relatively easy. Just go down your checklist and figure out what each is going to cost. The important thing is to find every direct overhead item in the job and put a cost by it. An estimate off by 20 or 30% isn't going to bankrupt you. But omitting any item is always a 100% miss.

Indirect overhead is different. You have indirect overhead even when all work stops. These are administrative or office expenses that exist just because you're in business: accounting fees, advertising, automobiles, depreciation on vehicles and equipment, donations, dues and subscriptions, entertaining, interest, legal fees, licenses, maintenance and repairs, office insurance, office phone, office rent, office salaries and benefits, office supplies, office utilities, payroll taxes, pensions, postage, profit sharing, small tools, taxes (property and business income), travel and uncollectibles.

Estimating indirect overhead is harder. The best way is to make two guesses. First, what's the cost of administrative overhead for the coming year? Either you or your accountant should have a pretty good guess on this figure. If you're in doubt, go back through your check stubs for the previous year. Add up all indirect overhead costs and then adjust that up or down to reflect higher or lower costs for the current year. Second, guess the total volume of business you'll have for the coming year. Be realistic. When you have these two numbers, divide the indirect overhead estimate by the gross volume estimate. If you're like most contractors, you'll get a number between 12 and 18%. Subcontractors may run a few percent less and remodelers and insurance repair contractors may be a few percent more. When you have the percentage, convert it to a factor that's easy to use when estimating: Subtract the percentage from 100%, divide the result into 1.00 and then subtract 1.00. That's the percentage to use on estimates when adding for indirect overhead. Let's look at an example.

Suppose you expect indirect overhead to be 15% of gross for the coming year. Subtracting that from 100%, you get 85%. Then divide 85% into 1.00 to get 1.176. Subtract 1.00 and you have 0.176. If your estimate of all job costs and job overhead, taxes and profit come to $100,000 on a particular project, multiply by 0.176 to estimate the indirect overhead cost of $17,600. Your bid would be $117,600. 15% of $117,600 is $17,640, the amount needed to cover indirect overhead.

Financing and Carrying Your Client

Few small volume builders have enough cash reserves to carry a construction project very far without progress payments from the client. If you're like most builders, your finances are so thin that clients usually have to carry you financially. Some clients don't mind advancing cash on a project. Others are willing to do it but make you pay for the privilege.

A client who meets your accelerated payment schedule probably wants to pay the lowest possible price. That type of work doesn't generate enough profit. Without adequate profits, you can't build cash reserves, forcing you to take more low profit work.

Profits are the lifeblood of a construction company. You take big risks as a builder. You have big expenses. It's only fair that the potential reward should be as big as the risk. You need adequate profits. Without them, you're stuck in low profit work. It's just a matter of time until a miscalculation or a slump brings a financial crisis.

Profits finance growth. They allow you to tackle larger and more profitable projects. The larger the project, the greater the potential reward and the more cash you need in your reserve.

Sometimes you'll have to carry the job and your client longer than expected. Every client has his own idiosyncrasies. Some pay faster than others. It's hard to tell in advance which clients will pay on presentation of an invoice and which will pay 30, 60 or 90 days later. Even the best of clients with the deepest pockets don't always pay promptly or in the amount expected.

Even if a client has the cash and is willing to pay, sometimes you're not going to get paid when expected. For example, a shortage of some key material, a labor dispute, or a string of weather delays can push back the payment date by weeks or even a month.

Some clients will intentionally try to get you behind the eight ball financially. Clients love to work on your money. It saves them from paying interest on their loans. It's up to you to keep clients on a short leash. Don't carry a client for more than 10% of the job cost or for more than 90 days. A client who needs more support than that is using you as a bank. Don't get into the banking business with any client.

Watch out for clients who play the leverage game. They'll get you in the hole for $50,000 or so and keep you there until a year after completion. This puts them in the driver's seat if repairs are needed during the guarantee period. They can get you back to fix things if they're holding a big chunk of your money.

The fact that a client delays payment for 30 to 60 days shouldn't create panic in your office if the amount due doesn't exceed about 5%. But holding 10% or more for 90 days should set bells ringing.

One exception is the client who delays payment until January 1st or July 1st to get the charge in a new fiscal year. I've permitted this, providing the client informs me beforehand and makes payment promptly when promised.

Since there's no way to avoid carrying a client occasionally, you should have an emergency plan to roll out when the time arrives. If you suspect a payment problem, bring your banker into the picture early. Bankers hate surprises and like to take their time in evaluating loan applications. Make sure your contract gives you the right to collect interest on unpaid balances and provides reimbursement for legal fees if collection is necessary. Discuss these charges with your client before sending the bill. These costs are a part of doing business. Your client should be prepared to pay.

Many construction companies arrange lines of credit with a bank. A credit line is like a preapproved loan. Get the line of credit approved first. Borrow when you need to and only as much as you need. Repay the loan and you re-establish the credit line at the original amount. The interest rate will usually vary as the prime rate varies.

Banks love a contractor who gets a job, borrows to finance it, finishes the job on time, collects, pays his loan off, and gets on to the next job. A builder who can do that regularly is considered a good customer at any bank.

Schedules

There are only two construction schedules that make sense. They're "Fast" and "Faster." Letting a job drag on longer than necessary runs up construction costs, increases losses due to weather, encourages pilferage, forces the transfer of crews and equipment between jobs, and makes collection more difficult.

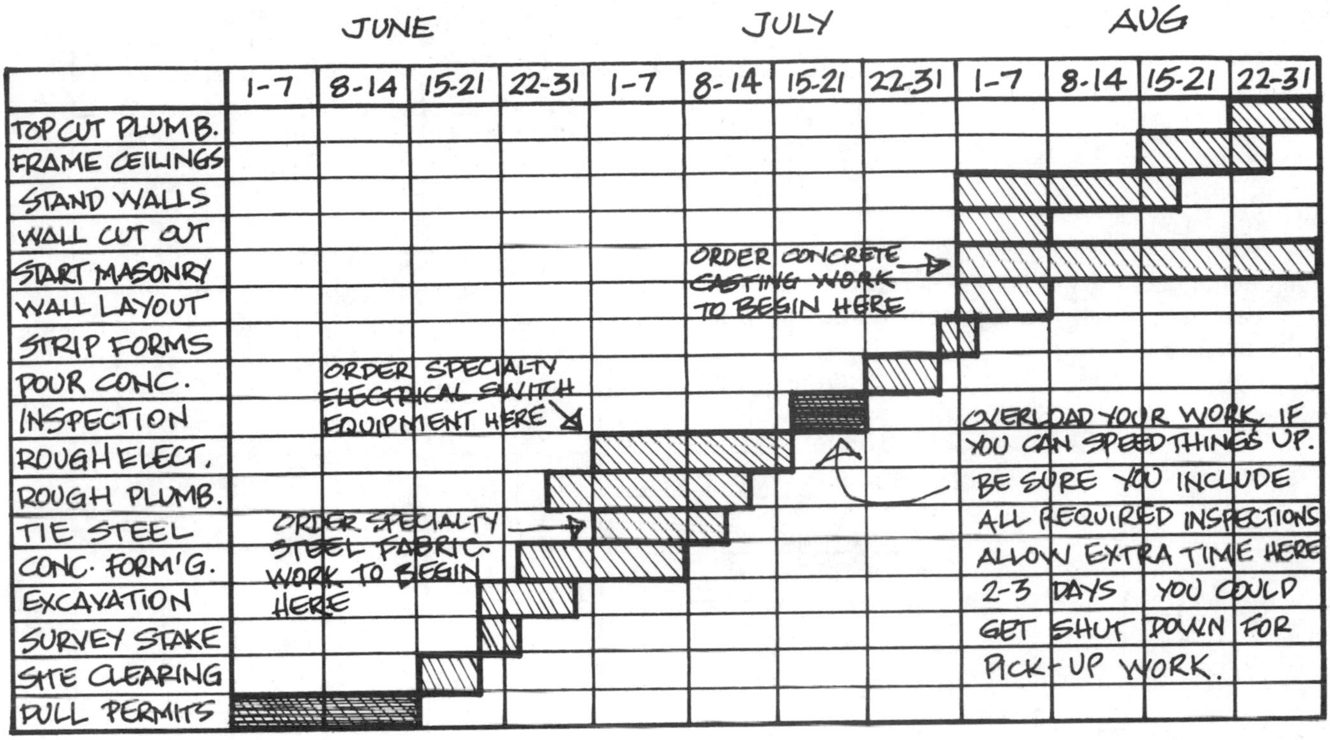

	JUNE				JULY				AUG			
	1-7	8-14	15-21	22-31	1-7	8-14	15-21	22-31	1-7	8-14	15-21	22-31
TOP CUT PLUMB.												
FRAME CEILINGS												
STAND WALLS												
WALL CUT OUT												
START MASONRY												
WALL LAYOUT												
STRIP FORMS												
POUR CONC.												
INSPECTION												
ROUGH ELECT.												
ROUGH PLUMB.												
TIE STEEL												
CONC. FORM'G.												
EXCAVATION												
SURVEY STAKE												
SITE CLEARING												
PULL PERMITS												

ORDER CONCRETE CASTING WORK TO BEGIN HERE

ORDER SPECIALTY ELECTRICAL SWITCH EQUIPMENT HERE

ORDER SPECIALTY STEEL FABRIC. WORK TO BEGIN HERE

OVERLOAD YOUR WORK IF YOU CAN SPEED THINGS UP. BE SURE YOU INCLUDE ALL REQUIRED INSPECTIONS. ALLOW EXTRA TIME HERE 2-3 DAYS YOU COULD GET SHUT DOWN FOR PICK-UP WORK.

Typical schedule
Figure 13-3

A well-controlled effort is the primary characteristic of a fast-paced contractor. Each subcontractor is notified of the start and completion dates and has an incentive to meet those dates. Bonus and penalty clauses encourage prompt performance.

A fast-paced schedule should include specific building department inspections and certifications. The schedule should consider the owner's needs, advantages of early building occupancy and special equipment installations. Figure 13-3 shows a three-month schedule from start to finish.

The only thing a fast schedule doesn't fully address is the early ordering of equipment and materials. It assumes that the job is awarded to the lowest bidder and that time isn't truly critical. Equipment and materials can be purchased and shipped as they are needed.

Scheduling a job where time is truly critical is usually done by computer. It shows the shortest distance between two points. In this case the two points aren't places; they're construction events. A fast-track schedule shows which tasks must be done first, second and third, which materials have to be ordered at which time, and when delivery is required. See Figure 13-4.

A fast-track schedule has one major advantage. It gives the owner and builder the opportunity to order materials and complete certain work before the final construction contract is signed and before the plans are approved by the building department. Pre-ordering can substantially reduce project duration, especially if major pieces of equipment have to be custom-built for the job. But it drives architects and engineers crazy. They have to set their final designs early and prepare several smaller sets of drawings for grading, steel, plumbing, sewer and power distribution. It's far more work for them, but it can be a definite advantage to everyone else.

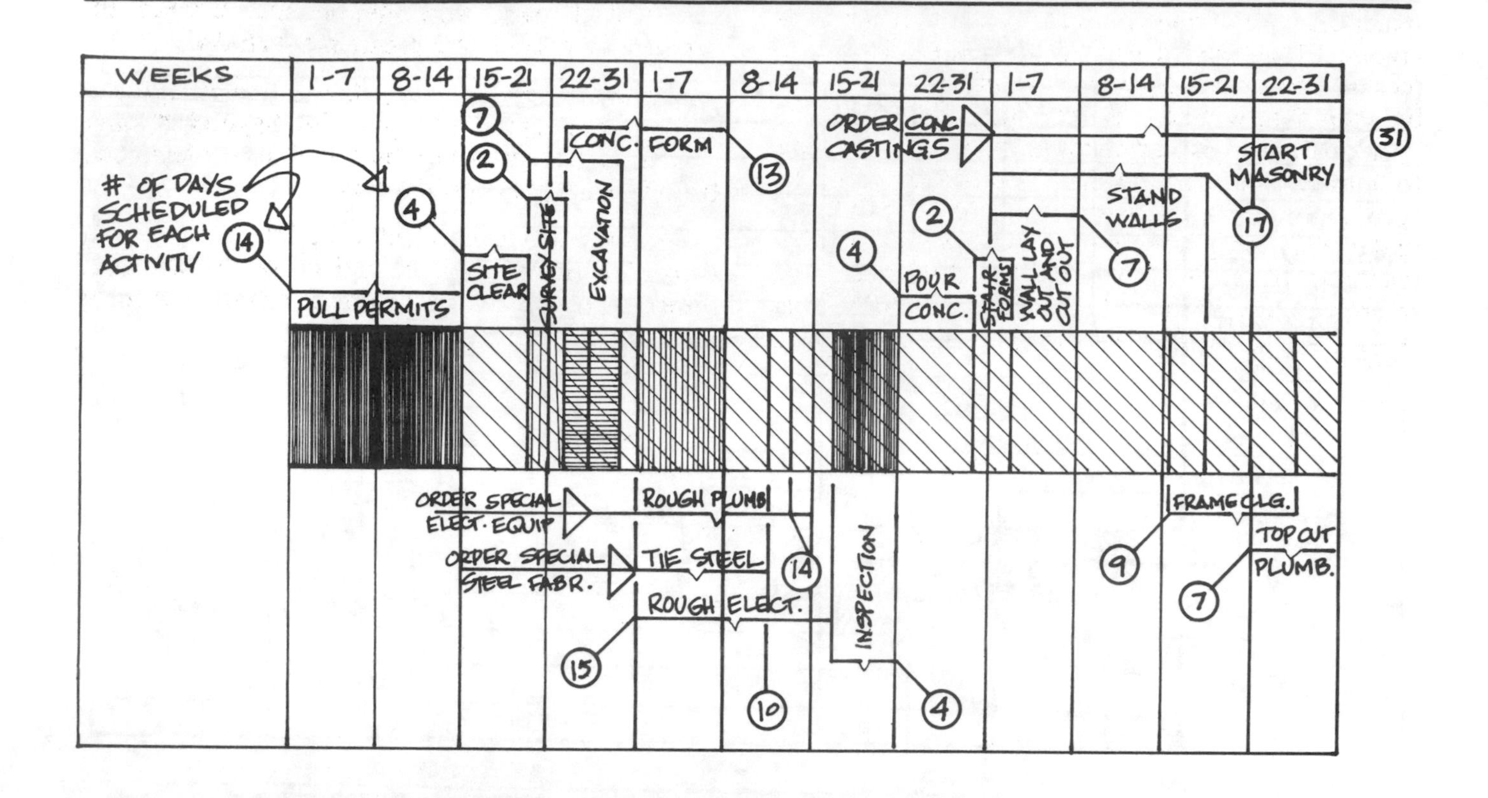

Fast-track schedule
Figure 13-4

Under the fast-track schedule method, the architects and engineers prepare a grading and site preparation plan that stands on its own without reference to any other drawings or documents. This plan is released to the owner and the contractor before the other plans are ready for bidding or checking.

The foundation plan is prepared next. It's a separate project set that can be scheduled and checked without referring to any other part of the job. Once the foundation plan is approved and construction begins, there's no turning back. All of the building must now conform to the foundation shape and maximum design weight. Changes made after the foundation is complete become expensive changes indeed.

Contracts

A contract should spell out the responsibilities of each party. Avoid surprises and you'll stay out of court. Be sure the contract covers the important points: what's to be done, by whom and when.

Every contract should have a start and finish date. When do you begin and when do you plan to finish? What happens if you finish early or late? What are the incentives and penalties? Naturally, you excuse delay due to weather and other acts of God. If you haven't considered all of these things, you don't have a very complete contract.

Set up a procedure for handling extras. Don't specify that the cost of extras will be "cost plus 10%." Instead, get the "cost of labor, materials, equipment, overhead and the contractor's usual profit" for extras.

Be sure to incorporate the plans and specs into the contract by referring to them expressly. And include the legal description of the property in the contract. You're encumbering legal title of the property with your work. If you have to sue, you want to cloud title to the right parcel.

Summary

For builders who have to survive on low bids, construction contracting is a real jungle with few survivors. Either find a way to negotiate jobs at set fees or find a specialty with less competition. Don't bid jobs that will have dozens of bidders.

If you're not skilled at estimating, take the time to master this essential art. You're not going to turn over pricing decisions to a hired hand. So make estimating your strong suit. Many schools teach construction estimating and several helpful references are listed at the back of this book. But there's no substitute for doing it the hard way: identifying every part of the project and then finding the material, labor, subcontract and equipment cost of each part. And then be sure all taxes, insurance, overhead and profit are included.

Remember my rule of thumb for carrying a client: Never carry a balance due beyond 90 days and never advance more than 10% of project cost. Beyond that only fools venture.

Finally, run projects at a sprint. Start a job, work like mad, finish it, and go on to the next. Finish three jobs, working one at a time, not three jobs working three at a time.

Investing in Inflation

As a construction contractor, there are two ways to measure your success. First, do you get personal satisfaction from your work? Second, are you making good money as a contractor? Sometimes the answer to the second question influences the answer to the first. If you're making a good living, you can expect personal satisfaction. That's just one reason to keep an eye on profits. Here's another. Profits tend to breed more profits. Many contractors multiply their profits by investing in inflation and betting on appreciation. This chapter is intended to explain how.

Profits give you money to invest. The more cash you have on hand at the end of each project, the more you can participate in the investment game. Spread your profits across a well-chosen piece of real estate, mix in a stiff shot of inflation or appreciation, and the result can be pure magic.

Many builders make good money in construction. But the most successful construction contractors also make money by investing in land and buildings — something builders tend to know intimately and can evaluate accurately. If you're interested in building net worth, your profit as a builder serves two purposes. First, it's a source of excess cash that can be used for investment. Second, it's a stream of future income that qualifies you as a lender when buying or developing more property.

As we discussed in an earlier chapter, *the real money in construction is made by project owners.* Look at Figures 10-2 and 10-3 back in Chapter 10. What's happening here? It's money working, invested capital, not the time or labor of the investor. That's why failing to make a profit hurts twice. Once for the cash you didn't make and once for the investment opportunities you had to pass up.

Profits are your ticket to the high-stakes inflation and appreciation game. If you're not making extra cash in construction, you're missing the boat.

While inflation may not go on forever, it's been very persistent for the last 40 years. The next 40 may be the same. Maybe it's built into the economic or political system in this country or the world. Maybe it's the result of some cause we don't understand. Maybe it doesn't matter. The important thing is that it exists.

Sure, inflation may ratchet down to 5% or less during some periods. But it's always cycled back up to higher levels within a few years.

Buying and Investing in Property

Here's my advice on buying land and planning projects. Don't buy anything unless you have the cash to develop it and know exactly how it should be developed. Don't buy raw land unless you plan to build on it soon. Holding raw land has no tax advantage beyond the interest you pay. Don't buy properties with heavily negative cash flow. You have enough drain on your time, energy and income without supporting underproductive property. Buy cheap but well-located property. Look for buyers anxious to sell. Make five offers at 10 or 20% below the asking price and one will probably be accepted. Buy because the price is right, not because you love the property.

As a builder, you know what property is worth and which areas can be developed profitably. Evaluate the property you build on. You know what the asking price of the land was. You know the cost of construction. It's easy to figure the owner's profit on a sale. Where owners are making good money developing property for sale or lease, you could do the same. You're in a position to be one of the most knowledgeable people in your area on land values and spec building opportunities.

If there aren't real opportunities in your back yard, by all means look elsewhere. Use the services of a realtor if necessary. But that will usually increase the cost by the realtor's fee. Whenever possible, find your own property. Keep the profits in your pocket. Don't pay others to do what you can do yourself.

Trading Property

Most property that changes hands today will involve a buyer and a seller. But experienced real estate professionals and many knowledgeable investors have discovered the advantages of trading real estate rather than selling it outright. During recession, trading can become the norm rather than the exception for many builders and realtors. When done correctly, trading avoids capital gains taxes while still allowing the new owner to begin depreciation of the property at the new value.

There are two basic kinds of trades in the real estate industry today. Sweat equity trading is by far the most common and easiest to transact. The second is known as the tax-free exchange. Let's look at sweat equity trades first.

Equity is the value of the property after all loans or liens have been paid off. If a home is worth $100,000 and the existing loans total $90,000, the equity is $10,000. But if the owner remodels the basement, paints the exterior and adds a porch, his equity may increase to $20,000 because the home's value becomes $110,000. The $10,000 increase is the result of his effort, not a reduction in the loans outstanding. Hence the name *sweat equity.*

But sweat equity isn't limited to homeowners improving their own homes. Professional builders can get in the act too. It's building sweat equity when you trade services for an equity position in real estate. Here's an example. I've illustrated it in Figure 14-1.

A landowner in your community is planning to build a new car showroom on his lot and lease it to a dealer on a long-term lease. He's asked you to bid on the job. After reviewing all bids and changing the design several times, the owner realizes that he hasn't got enough cash to finish the project.

Figure 14-1 shows that the owner is roughly $90,000 short of having enough cash. At this point, one of several things will happen. Usually the project dies and never gets built. The owner could bring in another investor with $90,000 and give him a specific interest in the finished project. But there's another option. Suppose you could have made about $90,000 on this job if your bid had been accepted. You offer to reduce your price by $90,000 in return for a $90,000 interest in the showroom. If that's acceptable to the bank, it may be an excellent deal for you. You receive a share of the rent each month and cash out when the property is sold.

A note of caution. Don't give away your overhead, too. Your office rent, phone, gas and utility bills still have to be paid. Invest your profit, but don't go beyond that. Cover all your costs, including overhead, contingency and a reasonable fee for your time. Investing more than your profit is the quickest way possible to go broke.

And don't invest $90,000 with anybody until you evaluate the stability and suitability of the primary owner as a partner. Check his credit, insist on references, talk with his attorney, investigate his banking connections, and review his audited financial statements for the past five years. Consult with your own attorney to make sure you haven't forgotten anything. If he checks out, take a chance. It's a good job that you wouldn't have otherwise, and it may become a good investment too.

This type of deal is only as good as the investment itself. Be sure the proposal is sound and that there's a way out. Your interest may be hard to sell by itself if the property isn't sold, so cover yourself with a buy-sell agreement or form a limited partnership that will actually own the property.

The second trading method is known as a tax-free exchange. It's a little more complicated. A tax-free exchange postpones income and thus postpones payment of taxes.

PROPOSED AUTOMOBILE DEALERSHIP BUILDING

1. COST OF LAND	$185,000
2. COST OF CONSTRUCTION	567,000
3. CARRYING COST THRU CONSTRUCTION	34,000
4. ARCHITECT & ENGINEERING FEES	29,000
5. OTHER MISC. COSTS TO DATE	8,000
6. LOAN COSTS	23,000
7. TOTAL COST OF PROJECT	$846,000
8. OWNERS CASH INTO PROJECT	−216,000
9. BALANCE BEFORE BANK LOAN	630,000
10. MAXIMUM BANK LOAN OBTAINABLE	540,000
11. REMAINING CASH REQUIRED OR DEAL FALLS APART	90,000
12. YOU AS CONTRACTOR PUT UP YOUR PROFITS	90,000
13. BALANCE REQUIRED	$ 0

NOW NOTE THAT YOUR INVESTMENT OF $90,000 HAS JUST BOUGHT YOU A 10.6% INTEREST IN THE PROJECT, AND AT A MINIMUM 5% INFLATION RATE YOUR $90,000 INVESTMENT WILL GROW TO $146,600. AN INCREASE OF $56,000 IN 10 YEARS. IF YOU DIVIDE $56,600 BY YOUR $90,000 INVESTMENT, YOU'VE INCREASED YOUR ASSET BY 63% AND YOU CAN'T LOSE MAKING A PROFIT. AND IF YOU DIDN'T LOAN YOUR PROFITS YOU WOULD NOT ONLY HAVE LOST THE $56,600 GROWTH BUT THE ORIGINAL $90,000 PROFIT AS WELL FOR A TOTAL LOSS OF $146,600.

Profit investment
Figure 14-1

A tax-free exchange is like any other purchase and sale except that no money changes hands. Here's how it works.

Let's say that you've owned an apartment building for several years. It's appreciated in value and your equity has grown considerably. But now you want to sell for two reasons. First, the annual tax deduction for depreciation and interest is a small fraction of the deduction available when the building was new. Second, you have the cash available now to buy a larger building.

But your accountant calculates that you'll owe a big chunk of capital gains tax if you sell now. You want to sell, but you don't want to hand the profits to the IRS. Consider a trade. It will both preserve your profit and delay the tax. Here's how it's done.

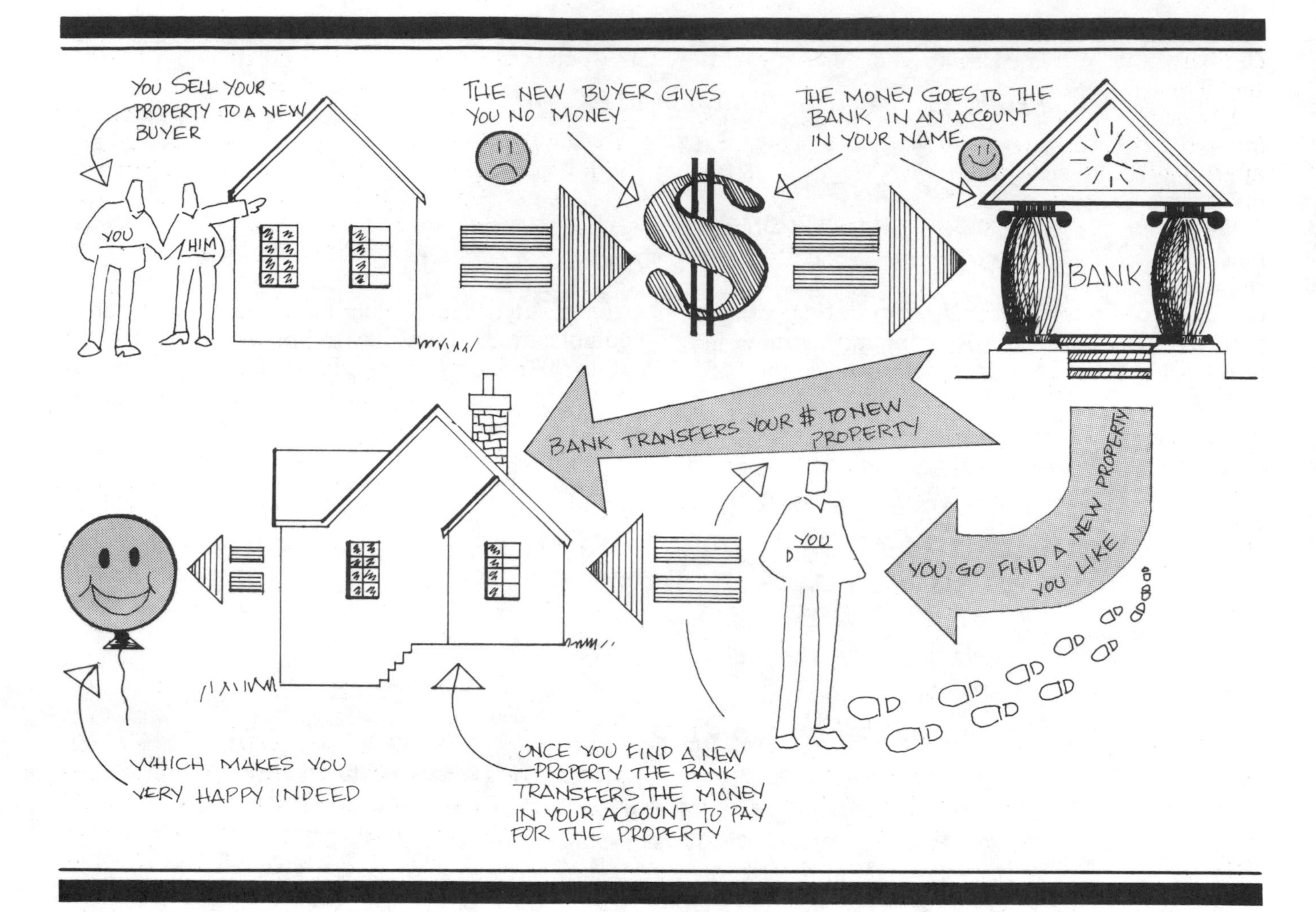

Trading real estate
Figure 14-2

First, find a willing buyer for your apartment building. Open an escrow or sign a sales agreement with that buyer. Now, go shopping for that new building. That's right — you have to become a buyer as well if you want to save on taxes.

Shop around for an apartment building you'd like to own. Inform your buyer of the property you choose. Your buyer then contracts to buy the property you located. At this point, two pieces of real estate have been tied up: the one you own and the one you would like to own.

Now the magic takes place. Your buyer uses his cash to buy the property you want. Next, he instructs the escrow company to trade the newly-purchased property to you for the property you already own, thus completing the transaction.

What you've got is a simple trade. You traded your real estate for your buyer's real estate. Figure 14-2 outlines the procedure. Keep in mind that the trade doesn't have to happen immediately. It can be postponed for a year or more, provided you don't actually get your hands on the money. Have the buyer put his purchase money in a savings account pledged in your name. Use these funds to buy your trade property once you've located what you like.

Once you own the new building, depreciation begins all over again. That's a big advantage if the property you traded away had been fully depreciated. Also, the exchange may involve a refinancing of the property, giving you spendable cash when the trade is complete.

On the negative side, tax-free exchanges are cumbersome, time-consuming and may be expensive if several real estate agents are involved.

Let me emphasize that tax-free exchanges *must involve like properties,* (apartment building for apartment building, land for land, etc.) and always involve an increase in value.

Of course, the capital gains tax will have to be paid eventually. But postponing the tax until you retire and drop into a lower tax bracket will usually cut the tax bite considerably. Also, the tax dollars you save now continue to compound, earning interest for you until they're finally paid to the IRS.

Deal Structuring and Financing

There must be ten thousand ways to structure a real estate deal. But no matter how you cut it, there's one thing to keep in mind. First and most obvious, the deal must be profitable for you. If it doesn't make sense on its face, drop it like a hot rock. Too many tax shelter deals turn out to be investment disasters because the investors consider only the tax benefits. If there's no profit, there's no reason to invest, regardless of the tax advantages.

The next most important consideration is allowing a way out of the arrangement. Keeping a land deal together may be harder than keeping a marriage together. The average land transaction lasts less than five years. Every transaction should define how the deal will be dismantled if things don't go as planned. This dismantling clause should define how profits, losses and cash are to be shared or dispersed. Otherwise the partners could be locked together indefinitely. Also, create a way for partners to sell out or buy out other partners.

The unwritten rule of premature partnership dissolution is that the partner leaving early carries the burden of his early exit. This encourages partners to see the partnership through to its original goal.

Once you've created ways into a deal, pay attention to the real meat of the transaction. This is pretty much a give-and-take proposition. Be sure to cover income, expenses, taxes, deductions, accounting, bank accounts and records. Once you've covered these issues, it's time to see your attorney, accountant, and insurance agent.

My best contracts have been verbal deals sealed with a handshake. The total agreement is usually something like, "If you finish it by that date, I'll pay you in full on completion." This kind of deal tends to be hassle free. But it isn't suitable for most land transactions.

The worst deals I've made have a common characteristic: bad financing. Every good deal is based on cash commitments that everyone can live with year after year if necessary. A bad financing package, a loan with payments you can't meet, has the potential of becoming a disaster for everyone.

You don't need to be a Philadelphia lawyer to get a good financial deal. The best financing is a simple, affordable loan. Fancy terms add fancy complications. Don't try to borrow more than the 80 to 90% most lenders will advance on residential property. On commercial property, you'll probably get less — 50 to 70%.

Where do you shop for financing? Some lenders are active and aggressive in certain areas and others are not. Lending policies vary at banks and savings and loan associations. Your best bet is a lender already lending on the type of project you're planning. Start with a lender who's familiar with your type of project.

Investing in Your Projects

When you've accumulated enough profits to begin developing your own land, do it right. Many really first-class speculative builders stub a toe by skimping on the wrong things. Believe it or not, the bulk of your profit or loss is sealed in at the planning stage of every project. Whatever you're trying to build or develop should be well-conceived, well-planned and well-executed. Let's look at the planning of a single-family spec home as an example.

Well-conceived starts with selection of the building site. It refers to your choice in the selling price range, and the look (eye appeal) of your finished project. It also refers to your choice of a design professional and selling agent. When you conceive of a project, plan its profitable outcome. If you ignore what your market can afford, what they want to buy, your costs, and your method of sales, you're setting yourself up for a loss.

Good planning requires forethought. Where are you going to spend a few extra dollars and where can you save a few bucks? These are important considerations.

Keep in mind that your job is to *organize, oversee and build.* You're the builder, not the designer. If it can't be done with a hammer, don't try it. Get professional help. It's easier to hire a good designer than it is to be one.

Good spec house design
Figure 14-3

If planning pays off, invest some money in a good plan. If you don't know which designer in your community is the best, all the better. In every community there are dozens of competent young designers and architects looking for a chance to do something special at a reasonable price. Ask several if they would like a shot at your project. Set up an informal competition. The winner will get the job. Hire three designers who are willing to work with you. Ask each to prepare a conceptual design. Explain the criteria for the job — size, area, site, price, materials, etc. Set a deadline for completion. Two weeks is enough. Have each quote a cost to prepare working drawings for their plan. Pay the losers $300 to $500 for his or her efforts. The winner gets to design your job.

By the way, this same competition works well for interior and landscape design.

But don't get carried away with design. Insist on a sensible and functional floor plan. There are about 25 proven, accepted floor plans. For each of these basic plans there are three or four simple variations. If you go much beyond this, you're gambling that your ideas are so red-hot that they just can't fail. My advice is to pick a standard floor plan and make only limited changes. Spend your money on finishes and decorative touches, not on some off-the-wall razzle-dazzle concept.

The exterior of your home should have clean, natural good looks. Don't spend a ton of cash on shutters, fancy fascias or used-brick planters. The basic shape, proportion and scale of your home should harmonize with other houses on the street and radiate a pleasant, comfortable feel by itself, even before you put on the interior and exterior finishes. The house in Figure 14-3 gives me this pleasant, comfortable feel.

Work for overall continuity in your home's exterior appearance. A quality home doesn't have 100% of the design goodies on the front and nothing on the other sides. Spread some of the design elements around your home. If you use brick or wood siding on the front, don't make it solid. Break it up. Put some on the sides and rear as well. The same with your stucco, paint and planters. Be consistent. Avoid curbside architecture.

Spend some time and money simplifying the design and avoiding waste. Build the simplest way possible. Keep dimensions in multiples of 2 feet to reduce waste of lumber, wallboard and cutting labor. Place one side of all windows and doors on the 2-foot module to reduce framing costs.

Keep the foundation plan simple. Adding an extra two exterior corners to a four-corner house will increase costs about 3%, even if nothing else changes and the floor area remains the same. A house with ten exterior corners will cost about 8% more than an otherwise identical house with four exterior corners. Align interior walls vertically to eliminate unnecessary framing. Align bathrooms vertically or place them back to back. Eliminate unneeded beams and custom-length roof joists and studs.

Spend enough to provide adequate space. Make the rooms big enough to live in and furnish comfortably. Give it enough bathrooms. Reduce or eliminate long hallways. Build a compact and efficient kitchen. Include a range, oven, dishwasher, garbage disposal, and microwave oven. Forget the trash compactor — it won't sell the kitchen. Consider a country kitchen that incorporates a breakfast table. No one needs a separate breakfast room. But provide a dining room big enough to seat eight if that's your market.

Increasing floor areas beyond the minimum makes good sense from two perspectives. First, it makes the house easier to sell. Second, homes with larger rooms are more likely to be over-appraised than under-appraised. Here's why. Small rooms are more expensive to build per square foot of floor than larger rooms. Small houses are more expensive to build per square foot of floor than larger houses. Most appraisers are not good judges of construction cost. They don't recognize the higher cost of small rooms and may not see the economies in larger rooms. Get a high appraisal and you increase the loan value. That makes it easier to get your selling price.

Children's bedrooms should be about 12' x 14'. The master bedroom can be 14' x 16' if there's a separate dressing area and bath. In the master bath, provide a tub and a double sink in the counter. Forget the gold fixtures and bidet. Avoid fancy wallpaper and harsh paint colors. Stick with muted tones that blend well.

Use masonry tile in the entry and consider hardwood for the dining room only. Use carpet with neutral colors elsewhere for a spacious, homey atmosphere.

Make your interiors tasteful and middle-of-the-road. Stay away from extreme design solutions. Don't let your wife be the interior decorator. I'll admit that wives are generally better at this than their husbands. But a trained professional can do a better job yet and is more likely to stay within a budget and take instructions gracefully.

Here are a few items that I always include in my spec homes. First, consider lighting. Don't put ceiling or wall lights in the bedrooms. Use switched wall plugs instead. Spend your lighting money to create a dazzling entry and an elegant dining room. Front porch lighting should cost at least $300. The dining room chandelier should run from $800 to $1200. Does that sound high? Try it. This is the voice of experience speaking. Spend the money and see what happens. If you think an 18''-diameter light seems about right, buy a 36'' light instead. Both the appraiser and potential buyers will remember an oversized fixture.

Spend a few extra dollars on a real masonry fireplace, not one of those cheap metal jobs. Don't build corner fireplaces. You can't sit more than two people in front of them. Favor vaulted open beam ceilings in the living and dining room. But make them simple to build. Don't use complex framing or hip-roofed ceilings.

Attention to basics sells homes. But the basics have to be beefed up and well presented. When the buyer and appraiser walk up to the front door of your new home, they should be able to sense and smell *quality*. But a feeling of quality doesn't have to cost bundle of money. Keep your costs down by keeping your construction as simple as possible.

Confuse the appraiser with the utter simplicity of your project, the scale of your ceilings, lighting and rooms, and the quality of your finishing touches. The selling price and loan value are based on the impression of value, not in the cost of construction.

In Summary

There's always room for one more successful builder in the construction industry. It might as well be you. But I know of no way to make it in construction without getting your house in order first. Spend some time thinking about your purpose and goals. There's no point in tackling big projects with only half-baked ideas. That's like setting out on a trip without a road map.

Be realistic about where you're going and what you plan to accomplish. Determine exactly what kind of work you want to do and the kind of company you want to develop. Find your niche in the construction business. Work at specialties that have a good profit margin. Avoid high risk jobs that could be financially terminal.

Consider what sacrifices you and your family are willing to make. Building is demanding work, make no mistake about it. It puts pressure on you, your wife and your children. But don't lose your perspective. Professional recognition is nice, but it doesn't replace the personal satisfaction that a family can provide.

Pay attention to swings in inflation. Every builder's future is tied to ups and downs in the economy. Inflation will give you a big financial boost if you own land and buildings that appreciate faster than the dollar declines in value. Use the multiplying effect of inflation to line your pockets with profits.

But there are times when it's simply better not to operate as a speculative builder. When times get bad, work on contract for government agencies or large corporations. Let *them* pay the bills and take the risks.

Keep some liquid reserves at all times. Cash is king. In good times and bad, money talks. Cash will keep you in the game when your competition is strangling. Have $5,000 to $10,000 available at all times, above and beyond your monthly expenses and investment commitments.

Make yourself available to your clients and creditors. Keep them informed and you'll keep them on your side. Use a company brochure to keep others informed of your company. Present yourself as an earnest, forthright and competent construction professional.

There's no way to avoid all the Flakes and Suede Shoe Operators who abound in the development business today. Your best protection is to get a financial statement, references, and find out early where the money's coming from. If you can't find the source of the financing, chances are that the project will never get built. If things don't add up, insist on more information until they do.

Local, state and national government are your partner on every project you have. Find some way to work in harmony. Don't get yourself into an adversary position with them. Promote a team effort. Keep the lines of communication open and flowing. Find out early what's required. Don't just wait for problems to come up. Take a leadership position. Start by getting on a first name basis with key people at City Hall, especially in the planning and building departments.

Every builder has to learn to compete in this business. Find out what work is available at what prices. Learn how to bid and how to avoid getting bid-peddled. You have to bid right to get significant work. If you're lousy at pricing, take the time and trouble to become an expert.

Remember, the real money in construction is in building for yourself. Developers make three profits: on the construction, on the land, and as an entrepreneur-speculator. In a strong economy, there's no better money-maker around. To make big bucks, you need a piece of the action, either through direct purchase, design-build, or trading. To make money, buy property, develop it and let inflation do the rest.

No one in any business has a better shot at building wealth than a knowledgeable, resourceful and aggressive construction contractor. If you're in the building business, you're in the right place to take advantage of the opportunities.

Good luck!

INDEX

Other Practical References

National Construction Estimator
Current building costs in dollars and cents for residential, commercial and industrial construction. Prices for every commonly used building material, and the proper labor cost associated with installation of the material. Everything figured out to give you the "in place" cost in seconds. Many time-saving rules of thumb, waste and coverage factors and estimating tables are included. **544 pages, 8½ x 11, $19.50. Revised annually.**

Building Cost Manual
Square foot costs for residential, commercial, industrial, and farm buildings. In a few minutes you work up a reliable budget estimate based on the actual materials and design features, area, shape, wall height, number of floors and support requirements. Most important, you include all the important variables that can make any building unique from a cost standpoint. **240 pages, 8½ x 11, $14.00. Revised annually**

Berger Building Cost File
Labor and material costs needed to estimate major projects: shopping centers and stores, hospitals, educational facilities, office complexes, industrial and institutional buildings, and housing projects. All cost estimates show both the manhours required and the typical crew needed so you can figure the price and schedule the work quickly and easily. **304 pages, 8½ x 11, $30.00. Revised annually**

Estimating Tables for Home Building
Produce accurate estimates in minutes for nearly any home or multi-family dwelling. This handy manual has the tables you need to find the quantity of materials and labor for most residential construction. Includes overhead and profit, how to develop unit costs for labor and materials and how to be sure you've considered every cost in the job. **336 pages, 8½ x 11, $21.50**

Cost Records for Construction Estimating
How to organize and use cost information from jobs just completed to make more accurate estimates in the future. Explains how to keep the cost records you need to reflect the time spent on each part of the job. Shows the best way to track costs for sitework, footing, foundations, framing, interior finish, siding and trim, masonry, and subcontract expense. Provides sample forms. **208 pages, 8½ x 11, $15.75**

Construction Estimating Reference Data
Collected in this single volume are the building estimator's 300 most useful estimating reference tables. Labor requirements for nearly every type of construction are included: site work, concrete work, masonry, steel, carpentry, thermal & moisture protection, doors and windows, finishes, mechanical and electrical. Each section explains in detail the work being estimated and gives the appropriate crew size and equipment needed. **368 pages, 11 x 8½, $26.00**

Builder's Guide to Construction Financing
Explains how and where to borrow the money to buy land and build homes and apartments: conventional loan sources, loan brokers, private lenders, purchase money loans, and federally insured loans. How to shop for financing, get the valuation you need, comply with lending requirements, and handle liens. **304 pages, 5½ x 8½, $15.25**

Spec Builder's Guide
Explains how to plan and build a home, control your construction costs, and then sell the house at a price that earns a decent return on the time and money you've invested. Includes professional tips to ensure success as a spec builder: how government statistics help you judge the housing market, cutting costs at every opportunity without sacrificing quality, and taking advantage of construction cycles. Every chapter includes checklists, diagrams, charts, figures, and estimating tables. **448 pages, 8½ x 11, $24.00**

Running Your Remodeling Business
Everything you need to know about operating a remodeling business, from making your first sale to insuring your profits: how to advertise, write up a contract, estimate, schedule your jobs, arrange financing (for both you and your customers), and when and how to expand your business. Explains what you need to know about insurance, bonds, and liens, and how to collect the money you've earned. Includes sample business forms for your use. **272 pages, 8½ x 11, $21.00**

Estimating Home Building Costs
Estimate every phase of residential construction from site costs to the profit margin you should include in your bid. Shows how to keep track of manhours and make accurate labor cost estimates for footings, foundations, framing and sheathing finishes, electrical, plumbing and more. Explains the work being estimated and provides sample cost estimate worksheets with complete instructions for each job phase. **320 pages, 5½ x 8½, $17.00**

Contractor's Growth and Profit Guide
Step-by-step instructions for planning growth and prosperity in a construction contracting or subcontracting company. Explains how to prepare a business plan: selecting reasonable goals, drafting a market expansion plan, making income forecasts and expense budgets, and projecting cash flow. Here you will learn everything required by most lenders and investors, as well as solid knowledge for better organizing your business. **336 pages, 5½ x 8½, $19.00**

Contractor's Guide to the Building Code
Explains in plain English exactly what the Uniform Building Code requires and shows how to design and construct residential and light commercial buildings that will pass inspection the first time. Suggests how to work with the inspector to minimize construction costs, what common building short cuts are likely to be cited, and where exceptions are granted. **312 pages, 5½ x 8½, $16.25**

Builder's Guide to Accounting Revised
Step-by-step, easy to follow guidelines for setting up and maintaining an efficient record keeping system for your building business. Not a book of theory, this practical, newly-revised guide to all accounting methods shows how to meet state and federal accounting requirements, including new depreciation rules, and explains what the tax reform act of 1986 can mean to your business. Full of charts, diagrams, blank forms, simple directions and examples. **304 pages, 8½ x 11, $17.25**

Bookkeeping for Builders
This book will show you simple, practical instructions for setting up and keeping accurate records — with a minimum of effort and frustration. Shows how to set up the essentials of a record keeping system: the payment journal, income journal, general journal, records for fixed assets, accounts receivable, payables and purchases, petty cash, and job costs. You'll be able to keep the records required by the I.R.S., as well as accurate and organized business records for your own use. **208 pages, 8½ x 11, $19.75**

Blueprint Reading for the Building Trades
How to read and understand construction documents, blueprints, and schedules. Includes layouts of structural, mechanical and electrical drawings, how to interpret sectional views, how to follow diagrams; plumbing, HVAC and schematics, and common problems experienced in interpreting construction specifications. This book is your course for understanding and following construction documents. **192 pages, 5½ x 8½, $11.25**

Builder's Office Manual, Revised
Explains how to create routine ways of doing all the things that must be done in every construction office — in the minimum time, at the lowest cost, and with the least supervision possible: Organizing the office space, establishing effective procedures and forms, setting priorities and goals, finding and keeping an effective staff, getting the most from your record-keeping system (whether manual or computerized). Loaded with practical tips, charts and sample forms for your use. **192 pages, 8½ x 11, $15.50**

Paint Contractor's Manual
How to start and run a profitable paint contracting company: getting set up and organized to handle volume work, avoiding the mistakes most painters make, getting top production from your crews and the most value from your advertising dollar. Shows how to estimate all prep and painting. Loaded with manhour estimates, sample forms, contracts, charts, tables and examples you can use. **224 pages, 8½ x 11, $19.25**

How to Sell Remodeling
Proven, effective sales methods for repair and remodeling contractors: finding qualified leads, making the sales call, identifying what your prospects really need, pricing the job, arranging financing, and closing the sale. Explains how to organize and staff a sales team, how to bring in the work to keep your crews busy and your business growing, and much more. Includes blank forms, tables, and charts. **240 pages, 8½ x 11, $17.50**

Manual of Professional Remodeling
This is the practical manual of professional remodeling written by an experienced and successful remodeling contractor. Shows how to evaluate a job and avoid 30-minute jobs that take all day, what to fix and what to leave alone, and what to watch for in dealing with subcontractors. Includes chapters on calculating space requirements, repairing structural defects, remodeling kitchens, baths, walls and ceilings, doors and windows, floors, roofs, installing fireplaces and chimneys (including built-ins), skylights, and exterior siding. Includes blank forms, checklists, sample contracts, and proposals you can copy and use. **400 pages, 8½ x 11, $19.75**

Handbook of Construction Contracting Vol. 1 & 2
Volume 1: Everything you need to know to start and run your construction business; the pros and cons of each type of contracting, the records you'll need to keep, and how to read and understand house plans and specs to find any problems before the actual work begins. All aspects of construction are covered in detail, including all-weather wood foundations, practical math for the jobsite, and elementary surveying. **416 pages, 8½ x 11, $24.75**

Volume 2: Everything you need to know to keep your construction business profitable; different methods of estimating, keeping and controlling costs, estimating excavation, concrete, masonry, rough carpentry, roof covering, insulation, doors and windows, exterior finish, specialty finishes, scheduling work flow, managing workers, advertising and sales, spec building and land development and selecting the best legal structure for your business. **320 pages, 8½ x 11, $24.75**

Craftsman Book Company
6058 Corte del Cedro
P.O. Box 6500
Carlsbad, CA 92008

10 Day Money Back GUARANTEE

- ☐ 30.00 Berger Building Cost File
- ☐ 11.25 Blueprint Reading for Building Trades
- ☐ 19.75 Bookkeeping For Builders
- ☐ 17.25 Builder's Guide to Accounting Revised
- ☐ 15.25 Builder's Guide to Const. Financing
- ☐ 15.50 Builder's Office Manual Revised
- ☐ 14.00 Building Cost Manual
- ☐ 26.00 Const. Estimating Ref. Data
- ☐ 19.00 Contractor's Growth & Profit Guide
- ☐ 16.25 Contractor's Guide to the Building Code
- ☐ 15.75 Cost Records for Const. Estimating
- ☐ 17.00 Estimating Home Building Costs
- ☐ 21.50 Estimating Tables for Home Building
- ☐ 24.75 Handbook of Construction Contracting Vol. 1
- ☐ 24.75 Handbook of Construction Contracting Vol. 2
- ☐ 17.50 How to Sell Remodeling
- ☐ 19.75 Manual of Professional Remodeling
- ☐ 19.50 National Construction Estimator
- ☐ 19.25 Paint Contractor's Manual
- ☐ 21.00 Running Your Remodeling Business
- ☑ 24.00 Spec Builder's Guide
- ☐ 16.75 Contractor's Survival Manual

Name (Please print clearly) ______

Company ______

Address ______

City/State/Zip ______

Total enclosed ______ **(In California add 6% tax).**
If you prefer, use your
Use your ☑ **Visa** ☐ **MasterCard** ☐ **AmExp**
Card no. 4128 739 235 039
Expiration date 06/92 **Initial** ______

Mail Orders
We pay shipping when you use your charge card or when your check covers your order in full.

In a hurry?
We accept phone orders charged to your MasterCard, Visa or American Express. Call 1-800-829-8123

These books are tax deductible when used to improve or maintain your professional skill.